Elementary Methods of Placer Gold Mining

by William Staley

with an introduction by Kerby Jackson

Introduction

It has been years since the Idaho Bureau of Mines released his important publication "Elementary Methods of Placer Mining". First released in 1944, this important volume has now been out of print for this days and has been unavailable to the mining community since those days, with the exception of expensive original collector's copies and poorly produced digital editions.

It has often been said that *"gold is where you find it"*, but even beginning prospectors understand that their chances for finding something of value in the earth or in the streams of the Golden West are dramatically increased by going back to those places where gold and other minerals were once mined by our forerunners. Despite this, much of the contemporary information on local mining history that is currently available is mostly a result of mere local folklore and persistent rumors of major strikes, the details and facts of which, have long been distorted. Long gone are the old timers and with them, the days of first hand knowledge of the mines of the area and how they operated. Also long gone are most of their notes, their assay reports, their mine maps and personal scrapbooks, along with most of the surveys and reports that were performed for them by private and government geologists. Even published books such as this one are often retired to the local landfill or backyard burn pile by the descendents of those old timers and disappear at an alarming rate. Despite the fact that we live in the so-called "Information Age" where information is supposedly only the push of a button on a keyboard away, true insight into mining properties remains illusive and hard to come by, even to those of us who seek out this sort of information as if our lives depend upon it. Without this type of information readily available to the average independent miner, there is little hope that our metal mining industry will ever recover.

This important volume and others like it, are being presented in their entirety again, in the hope that the average prospector will no longer stumble through the overgrown hills and the tailing strewn creeks without being well informed enough to have a chance to succeed at his ventures.

Kerby Jackson
Josephine County, Oregon
October 2015

TABLE OF CONTENTS

LIST OF ILLUSTRATIONS

Moscow, Idaho
April 27, 1934

Dr. John W. Finch

Director, Idaho Bureau of Mines and Geology

Sir:

Material is submitted herewith for a pamphlet on placer mining methods for the more or less inexperienced prospector. The object throughout this discussion has been to present the material in as non-technical language as possible. This is thought to be necessary because of the inquiries received in the past in the Bureau office from so many who have not had technical training.

An elaborate discussion of the geogolical principles involved in the formation of gold placers is not attempted because it is not considered to be essential to the purpose of this paper.

No attempt has been made to give a complete bibliography of placer mining. Books or articles which are non-technical and easily understood by mining people make up the bulk of the list. Also, a point was made to include books easily obtainable. It is realized that a number of the government publications have been out of print for some years; however, they may be obtained from some libraries.

This paper is not meant to be a treatise on alluvial mining. The experienced miner or one of means is referred to the more extensive works on this subject.

Since the publication of the third edition of Pamphlet No. 35, the present edition has been enlarged to include a discussion on undercurrents as a means of removing fine gold, amalgamation procedure, operation of dry placers, and other minor details suggested by inquiries.

It is hoped that this paper may serve the prospector and contribute to the furthering of the mining industry in Idaho.

Respectfully yours,

W. W. STALEY

Mining Engineer, Idaho Bureau of Mines
and Geology

BRIEF HISTORY OF ALLUVIAL MINING

Alluvial mining is thought to be the oldest mining method. Records left by the ancients mention it as the means used for obtaining gold and silver. Of the many mining methods for obtaining valuable minerals, alluvial mining presents the least difficulties. There is very little, if any, drilling or blasting necessary. For this reason the early miners confined the greater part of their attention to alluvial mining. The people of ancient Egypt, many centuries before the birth of Christ, washed gold from the stream beds of the surrounding country*.

Most of the important gold-producing areas of the world were discovered because of placer operations. Among these may be mentioned California, Colorado, South Dakota, Idaho, Alaska, and the Yukon territory in America. South Africa and India are important foreign districts.

The early methods of extracting gold from the sands and gravels in which it occurred were confined to panning and the use of rather crude forms of rockers and sluice boxes.

GEOLOGY OF ALLUVIAL DEPOSITS

Placer deposits in Idaho are masses of loose gravel and sand, containing gold and other valuable minerals.

Substances Likely to Occur in Placers

While the word "placer" usually causes one to think of gold, it must be remembered that many other substances may be found in placers. Of importance among these are platinum, gems, silver, tungsten, tin, and minerals containing the rare metals.

Formation of Placers

The gold found in placers originally existed in place as deposits of various forms in areas intruded by igneous rocks. In some cases, it was deposited in the igneous rock itself in finely disseminated particles; in other cases, it other cases, it was originally in quartz veins, cutting through the igneous a n d o t h e r rocks and formed as a result of the igneous intrusions. Due to disintegrating processes (change of temperature, wind, rain, earth movements, and chemical action) the rock containing the gold has been reduced to such a state that it is easily broken and the gold freed. Through the action of running water and of glaciers in some instances, the gold-bearing rock is transported away from its source. The moving water causes the heavier gold particles to work slowly toward the bottom of the stream bed. On reaching bedrock, or hard pan, the gold moves slowly down stream until it lodges in crevices, cracks, or other irregular openings in the stream bed.

Placer deposits may be moved many times, depending upon the volume of water and the velocity with which it is flowing, and this generally depends upon the rising and subsiding of that particular part of the earth's crust. There is no fixed rule as to where the gold is apt to occur in the stream bed. The velocity of the stream is not the same at all points in its cross section. Points where the bed has widened, with resultant decrease in velocity, are the most favorable. The reason for this is that the gold is given the chance to settle to the bottom, when velocity of water decreases. Placers may be found in old dry stream beds. At the time of this formation, water was, of course, present. Later disturbances may have caused the stream to change its course. Or climatic conditions may have been responsible for its drying up.

---

* Lock, A. G., Gold: Its Occurence and Extration.

Wilson, E. B., Hydraulic and Placer Mining.

There have been very few instances. where the gold in a placer deposit has been traced back to its source. The reason is that the source has been either completely eroded away, or has been deeply covered with other material, such as lava flows, sediments, etc., or the gold may have traveled great distances. It has been rather definitely proven that there are cases where placed gold was found over one hundred miles from its original source.

Classification of Placers

A. H. Brooks* considers that there are three conditions operative in the formation of placers: (1) The occurrence of gold in bed rock to which erosion has access, (2) the separation of the gold from bed rock by weathering or abrasion, (3) the transportation, sorting, and deposition of the gold-bearing material derived by erosion. His statement is as follows:

"The distribution and origin of the gold in bed rock, involving as it does the study of ore deposits, although of first importance to the study of placers, can here be only briefly discussed. Of equal importance and more closely related to the genecis of placers in the consideration of the agencies leading to the separation, sorting and deposition . . . . In the text-books emphasis has usually been laid on the two types, the residual placer and the transported or true placers, without full recognition of the fact that the former often represents an intermediate stage between the bed rock source of the gold and the true placer.

The transportation, sorting, and deposition of material furnished by the weathering of rocks, the most easily understood of geologic phenomena, are all important agencies in placer formation. . .

A logical classification of the placers should be based, first, on genesis, second, on form. The primary grouping, according to origin, would be "residual placers," "sorted placers," and "re-sorted placers." The residual placers are those in which there has been no water transportation, the concentration of the gold being due solely to rock weathering. The gold of the sorted placers is the result of transportation, sorting,, and deposition by water. Placers of the third group are those in which the gold has passed through two or more cycles of erosion before its final deposition. Those of the first class are practically all of one type. The sorted and resorted placers embrace m a n y subordinate types, named according to the form of occurrence. The following list presents the larger groups and the more important of the subordinate types:

1. Residual placers.

2. Sorted placers.

    a. Hillside

    b. Creek and gulch

    c. River-bar

    d. Gravel plain

    e. Bench

    f. High bench

---

* The Gold Placers of Parts of Seward Peninsula, Alaska. U. S. Geological Survey Bull. 328.

3. Re-sorted placers.

    a. Creek and gulch

    b. **Beach**

    c. Elevated bench

A brief description of the more common types listed above follows*.

Residual Placers

These are placers in which the gold is accumulated in place by the disintegration of the rock containing it. It is not transported from its original source.

Hillside Placers

These are very old deposits, occurring on the tops and sides of hills. They may have been left in this elevated position because of earth disturbances which lifted the area above the former stream bed, or the original stream which deposited them may have changed its course or have meandered to a new bed.

Creek Placers

These are the best known and most productive placers. Brooks† has described this form of placer as follows:

"The pay streak in these deposits is usually on bed rock, though it sometimes is found on a clay which overlies the rock. Where no clay is present the gold is found not only on the bed rocks, but also where the rock is broken the gold has worked its way down into the joints and crevices. Streams are often found to have a layer of clay on bed rock, which gradually thins out up-stream and finally disappears entirely. The presence of the clay on bed rock usually indicates that no gold will be found in the weathered rock below, as the impervious layers prevent the gold from working its way down."

The entire width of the stream should be tested as the pay streaks are very irregular. They usually run parallel to the direction in which the water is flowing.

Gulch Placers

These are very similar to creek placers, except that there is now very little, if any, flowing water present.

River-bar Placers

These are bars of gold-bearing sand or gravel that have been laid down by large streams or rivers. The gold is usually distributed throughout the bar. There is often more fine (flour) gold than coarse. The deposits are usually very low-grade as compared to creek placers.

---

* The Gold Placers of Parts of Seward Peninsula, Alaska; Bull. 328, U. S. Geological Survey.

  Longridge, C. C., Hydraulic Mining; The Mining Jour. London.

† Brooks, A. H., Reconnaissance in the Cape Nome and Norton Bay Regions, Alaska; Special Publication, U. S. Geol. Survey, 1901, p. 140.

Bench Placers

These are more or less ancient placers, occurring in bench or terrace form, on sides of valleys or courses of ancient streams, from 50 to 300 feet or more above the present stream level. The presence of well rounded gravel is indicative of material carried and sorted by water. Figures 1 and 2 illustrate creek and bench placers.

Position of Gold in Deposit

In general, the various sized particles of gold or other placer minerals will be found in the following section of the stream when the water has flowed continuously in one direction:*

1. The coarse gold will be deposited in the upper part of the stream.

2. The finer gold will be deposited in the lower portions of the stream.

3. The richest and coarsest gold will be deposited in the layers of comparative coarse gravel wash.

4. The finer gold will be deposited in the finer sandy drifts.

5. The best gold should occur in the layers of wash containing black sand and pebbles of magnetite or other heavy mineral.

6. On a favorable bottom, gold will be ordinarily lodged on the down side of a bar of rock running across the bed of a stream.

Associated Minerals

Black sand (magnetite, an oxide of iron) is nearly always found in placers with gold. Its presence or absence is not positive proof of the presence or absence of gold. Ilmenite (an iron titanium oxide) resembles magnetite to a large extent. It is usually present. Garnet (ruby sand) and zircon commonly occur in gold placers. In Alaska†, and in at least one locality in Idaho, cinnabar (mercury sulphide) has been found in gold placers. Scheelite (calcium tungstate) and cassiterite (tin oxide) have been found in some places. Pyrite is commonly found, and by the inexperienced prospector may be confused with gold. A very simple test quickly distinguishes between the two. Pyrite is very brittle. A slight pressure between two hard surfaces reduces it to fragments. Gold is simply flattened without breaking. Biotite mica, which has altered to a bronze color, is sometimes confusing. It is readily told from gold by the readiness with which it breaks when bent back and forth.

SAMPLING OF PLACER DEPOSITS

Before any extensive operations are attempted, the placer deposit should be sampled. This, of course, applies where large scale sluicing, dredging, or hydraulicking, is contemplated, and not where the gold pan, rocker, or some such elementary process is used.

There are two general methods of sampling; test pits and bore holes. Test pits are most profitably used in shallow deposits (probably not deeper than 25 feet). For greater depths the churn drill should be used. The test pit or shaft gives a more accurate sample. It covers a larger area; the gold contained in the gravel is removed with the gravel with very little concentrating of gold as the bottom of the shaft is approached. With the churn drill it is difficult to prevent concentration

---

* Longridge, C. C., Hydraulic Mining, p. 12 (1910).

† The Gold Placers of Parts of Seward Peninsula, Alaska, Bull. 328, U.S.G.S.

of gold. The final choice between the two methods rests with the cost. If the shafts must be timbered, or water pumped out, the bore hole method may be far cheaper.

The material removed from the shaft or the bore hole is panned, sluiced, or amalgamated, to remove the gold. The gravel is weighed, or its weight calculated, from the size of the opening from which it came. The recovered gold is weighed and expressed in cents per cubic yard.

Care must be taken against "salting" the sample, i. e., getting gold into it that does not belong there.

To determine the grade of fineness of the gold, it will be necessary to send a sample to an assayer. Placer gold varies between about $17.00 worth of gold per ounce to almost pure gold, the present price of which is $35.00 per ounce. It is found nearly always alloyed with varying amounts of silver.

In sampling or working a deposit, one must be sure that he has reached the real bed rock before abandoning the claim. Figure 3 illustrates this. It will be noted that the gold has been deposited in alternate layers with clay. This indicates changing condition of deposition.

DESCRIPTIONS OF THE SIMPLER MINING METHODS AND APPARATUS

The size of the gold to be recovered has an important bearing on the details of the appliance to be used. Finely divided gold is much more difficult to save than the coarser variety.

The following table will give some idea of the size of gold particles and their values.*

Nuggets

Coarse gold—that which remains on a 10-mesh screen (ten openings per linear inch).

Medium gold—that which remains on a 20-mesh and passes a 10-mesh screen (about 2200 colors to 1 oz.)

Fine gold—that which passes a 20-mesh and remains on a 40-mesh screen (about 12,000 colors to 1 oz.)

Very fine gold—that which passes a 40-mesh screen (about 40,000 colors to 1 oz.)

Flour Gold

Purington quotes examples of finely divided gold:

170 colors to 1 cent (314,500 to 1 oz.).

280 colors to 1 cent (436,900 to 1 oz.).

500 colors to 1 cent (885,000 to 1 oz.).

Of the many methods that are used for recovering gold from placer deposits there are only three that merit description in so far as the prospector is concerned. In the order of simplicity, the construction of the apparatus and operation of these three methods follow.

---

* Young, G. J., Elements of Mining, 2nd Ed. (1923) p. 400.

Panning

The ordinary sheet-iron gold pan varies from about 10 to 18 inches in diameter at the top. The depth is about three inches. The ordinary 10-inch frying pan with the handle removed is quite often used. This pan holds about five pounds. The 18-inch pan holds about 25 pounds of dirt. Figure 4 illustrates the gold pan.

The gold pan is made of stiff sheet-iron. The inner surface must be kept clean and bright, and free of grease. Some pans are made with a copper bottom. Copper amalgamates readily with mercury. By rubbing mercury on the copper bottom, fine gold is retained through amalgamation.

Operation of Pan

The pan is filled about two-thirds full of dirt and placed under water. While in this position the contents are stirred or "kneaded" with both hands. This procedure is necessary to break up the lumps and to free the gold from clay-like material. As the disintegration proceeds, the large stones and pebbles are thrown out. When the material has been thoroughly broken up and the large rocks removed, the pan is taken in both hands for the panning operation. The position of the hands is slightly back of the middle of the pan. This permits the pan to be inclined down and away from the operator. The pan is now raised until it is just covered with water. It is now given a slight oscillating, circular motion, with the result that the contents are shaken from side to side. This motion keeps the lighter material in suspension and washes it out of the pan. It also enables the gold and heavy minerals (magnetite, etc.) to work their way to the bottom of the mass. This operation is continued until only the gold and black sands are left. This material is now dried and the magnetitie removed with a magnet. Other material, such as stream tin and heavy non-magnetic minerals, are separated from the gold either by amalgamating the gold or by picking out the gold, piece by piece. The separation of gold from the mercury used in amalgamation will be discussed later in this paper.

Peele*, Wilson†, and Longridge‡, state that about 100 pans of dirt are the most that can be panned by an experienced miner in 10 hours. Assuming that placer gravel weights 135 pounds per cubic foot, and that the gold pan holds 15 pounds, 100 pans would be equivalent to about 11 cubic feet or 4/10 of a cubic yard. With the large pan (18 inch diameter), a good panner may handle one cubic yard.

Rockers

There are many forms and sizes of rockers. The rocker handles about three to five cubic yards of material per 10 hours, its capacity depending upon the size of the gold and the amount of clay present. Large amounts of clay slow the operation down. It is necessary that all the clay be washed free of the gold, otherwise, the fine gold is floated away. The sketch shown as Figure 5 illustrates a convenient form of knockdown rocker.**

Description of rocker:

The inside of one side of the rocker and an end view of the rocker is shown.

A—Cleats for holding the back of the rocker.

---

* Peele, R., Mining Engineer's Handbook, 1st Ed., vol. I, p. 755 (1918)

† Wilson, E. B., Hydraulic and Placer Mining, 3rd Ed., p. 63 (1918)

‡ Longridge, C. C., Hydraulic Mining, p. 181, (1910) (The Mining Journal, London)

** Storms, W. H., How to Make a Rocker; Eng. & Min. Jour., June 24, 1911, p. 1243.

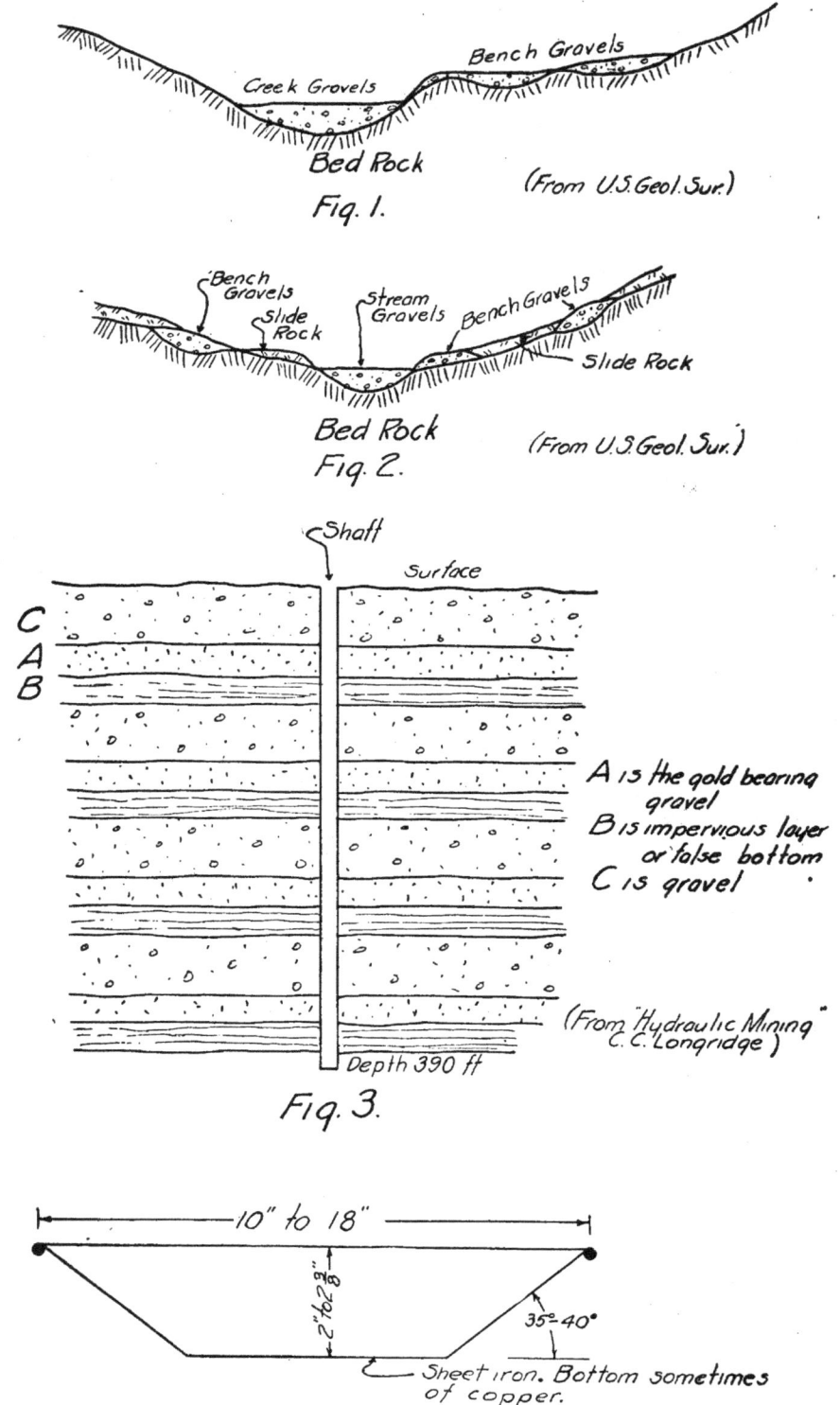

Creek Gravels

Bench Gravels

Bed Rock

Fig. 1.

(From U.S. Geol. Sur.)

Bench Gravels

Slide Rock

Stream Gravels

Bench Gravels

Slide Rock

Bed Rock

Fig. 2.

(From U.S. Geol. Sur.)

Shaft

Surface

C
A
B

A is the gold bearing gravel

B is impervious layer or false bottom

C is gravel

(From "Hydraulic Mining" C. C. Longridge)

Depth 390 ft

Fig. 3.

10" to 18"

2" to 2⅞"

35°-40°

Sheet iron. Bottom sometimes of copper.

Fig. 4.—Gold Pan

— 8 —

B—Cleat for holding bottom of rocker, L.

C—Cleats for holding back of rocker

D—Cleat for holding canvas apron frame.

E—Cleats for holding brace at top of rocker.

F—Cleat for holding sieve box.

X—Bolt holes for ½ inch iron bolts used in holding rocker together.

I—Riffles ¾ inch high by 1 inch wide.

K—Rockers.

H—Handle for rocking apparatus.

L—Bottom board of rocker.

M—Spike projecting 1½ inches to prevent rocker from slipping down grade.

The bottom board, L, of rocker should be in one piece. This is to prevent leakage of fine gold which might occur if two poorly fitted boards were used. Material of construction is preferably finished ¾ inch. The six ½ inch rods should have nuts and washers for the ends. This permits tearing the rocker down for transportation purposes.

The dimensions of the sieve box are as shown in the sketch. It should just fit loosely in the top of the rocker. The bottom is made of heavy sheet iron perforated with about ½ inch diameter holes.

The apron is a framework made of 1 inch by 1½ inch material well fitted together and covered with canvas. The canvas is not stretched tight, but allowed to sag somewhat at the bottom. This gives a slight depression in which gold is caught.

The grade or inclination of the rocker is obtained as follows:

Two heavy planks are firmly placed on the ground such a distance apart that each of the rockers will fall about in the center of a plank. The planks must have holes in them to receive the spike in the bottom of the rockers. The plank under the front or discharge end of the rocker is placed two inches lower than the rear plank. This arrangement, therefore, gives a drop of two inches in three feet. The grade is influenced directly by the following conditions:

1. Rapidity with which material can be fed to the rocker.

2. Amount of clay present.

3. Fineness of gold.

If the gravel is finely bound together with clay, the grade should not be less than two inches. If very little clay is present, and the gold is not too fine, the grade can be increased. In any event, the grade must be such that the clay is completely removed from the gold before the discharge is reached, and if the gold is very fine it should be given a chance to settle. In cases of very fine gold and considerable clay, it might be advisable to add one more riffle.

Operation of Rocker

For the operation of the rocker much more water is required than for the gold pan. Where there is a shortage of water, it is usually better to carry the gravel to a point near the source of water. The gravel is placed in the screen box and the rocker is shaken back and forth with a vigorous motion. At the same time, water is poured over the gravel, or a small stream of water is permitted to run over it. If

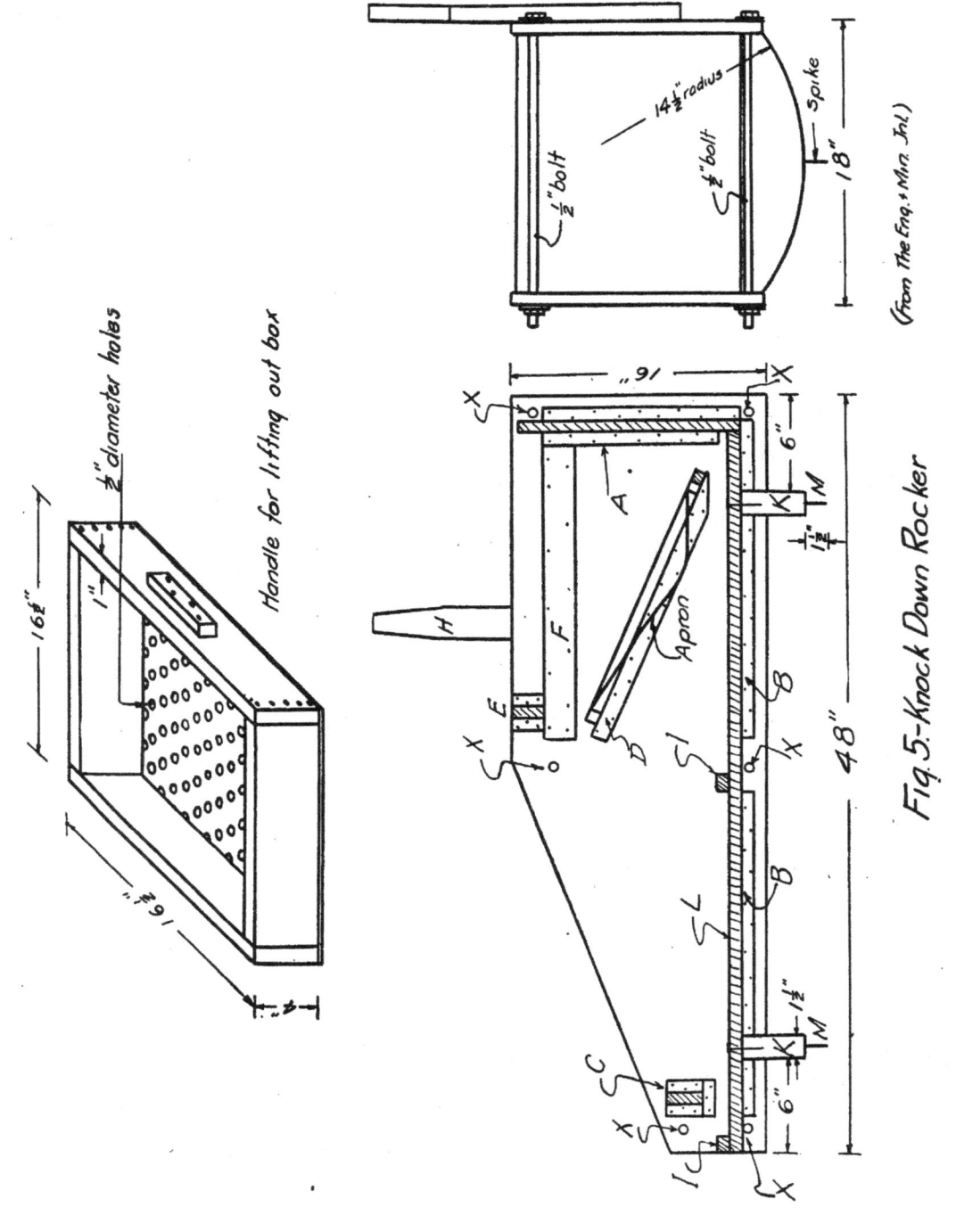

½" diameter holes

Handle for lifting out box

16⅜"

1"

16⅜"

3"

¼"bolt

14½"radius

¾"bolt

Spike

18"

(From The Eng. & Min. Jrl.)

16"

X

A

H

F

Apron

E

D

I

K

M

1½"

B

B

L

L

48"

6"

C

X

C

X

K

M

1½"

6"

Fig. 5-Knock Down Rocker

— 10 —

water is scarce, the discharge can be caught in a small pool and rinsed. Good judgment must be exercised in the use of water. If too rapid a flow is used, the smaller gold particles will be washed over the riffles and lost with the discharge. At the same time, sufficient water must be used to completely disintegrate the gravel and remove the clay. An attempt should be made to keep a fairly steady stream flowing rather than an intermittent, surging supply. The amount of water must be sufficient to carry the tailings over the riffles. The motion of rocking is a quick jerk with a sudden stopping of the motion. The heavy sands must not be permitted to build up back of the riffles. If this is allowed, the gold will wash over these sands and be lost.

Clean Up

The canvas or blanket forming the apron is rinsed off in a tub of water two or three times a shift. The gold and sands back of the riffles are removed as often as thought necessary. The concentrates are dried and the gold removed in the same manner described under panning.

The rocker is not very efficient. It permits the handling of more material than does the gold pan. Mercury some times is placed back of the riffles to catch some of the fine gold.

When the over-size material is removed from the sieve box, it should be inspected for nuggets before being discarded.

The tom or long tom is sometimes used in place of the rocker. It is illustrated in Figure 6. Six to twelve foot sluice boxes are used. One man shovels the gravel into the head box, others lift out boulders with pitchforks and break up the lumps of clay. Clean up is made in the same manner as for the rocker.

Sluices

In the use of sluice boxes two conditions may arise. First, where the box rests on the ground, the second, where it is necessary to elevate the sluice on trestles, necessitating also the elevating of the gravel. Only the first case will be discussed. The construction of the boxes and the manner of retaining the gold are the same in either case.

Material

The material from which the sluices is made is rough-finished lumber. There are some instances, such as dredging and large scale hydraulicking, where metal boxes are used. In many cases the box will be made of lumber which has been hewn out by the prospector himself.

Dimensions

The sluice is made up in sections. These sections vary from 12 to 16 feet in length, depending upon the locality. Twelve-foot sections are the most common. The width varies from one-foot to five feet, but is usually between 12 and 18 inches. The depth is from eight to ten inches. The boards from which the boxes are made are about one and one-half inches thick.

Construction

The boxes are made of rough lumber. For ordinary work the following dimensions are sufficient:

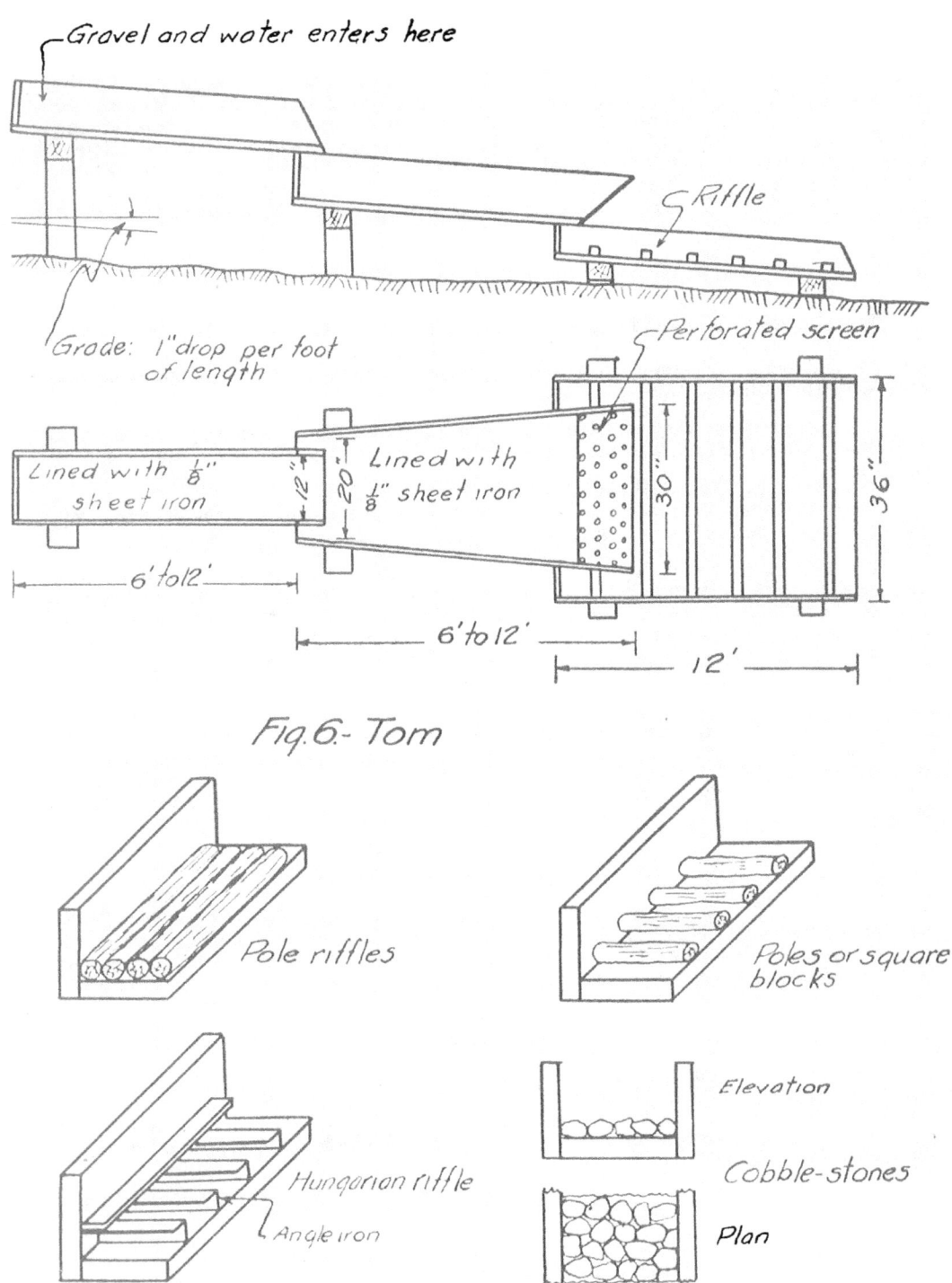

Gravel and water enters here

Riffle

Grade: 1" drop per foot of length

Perforated screen

Lined with ⅛" sheet iron

Lined with ⅛" sheet iron

12"

20"

30"

36"

6' to 12'

6' to 12'

12'

Fig. 6.- Tom

Pole riffles

Poles or square blocks

Hungarian riffle

Angle iron

Elevation

Cobble-stones

Plan

Fig. 7.- Various Types of Riffles

Length: 12 feet

Width: 1-foot inside measurement

Depth: 8 inches inside measurement

Thickness of material: 1½ inches

One end of each box should be narrower than the other. This permits the telescoping of the boxes. As the gravel bank recedes, boxes from the discharge end are brought to the head end. Thus, it is not necessary to move the entire sluice in order to keep close to the working face.

Head Box

The box in which the gravel is shoveled is called the "head box." It is equipped with a grizzly or bars to prevent the large boulders and rocks from entering the sluice. This is also where the water enters the sluice.

Grizzly

The grizzly is made of iron bars or heavy pipe. The spacing between the bars will depend upon the size of the gravel. If only medium sized gravel with very few large rocks to be encountered, a perforated sheet may be used.

Riffles

The riffles can be constructed of many different things: Wooden blocks, angle irons, poles, cobblestones, boulders, etc., have been used. They may run the length of the box or across it. Figure 7 shows some of the riffles in common use. The boxes are shown with one side removed.

In Figure 8 is shown a section of a sluice. The number of boxes making up the sluice depends upon the amount of material to be handled and the size of the gold. Fine gold requires more time to settle.

When real fine gold is present, the last sluice box may be replaced by a very wide table* (about 16 feet) from 10 to 20 feet in length. A screen is placed over the end of the sluice box so that only the sands and fine gold can get onto the table. The table is divided into sections eight feet wide and each half covered with burlap tightly stretched. The material is allowed to flow over one-half for about 12 hours. Then it is changed to the other side. The burlap is removed and washed off in a tub.

In some instances, mercury may be placed back of the riffles in the boxes near the discharge end of the sluice. This helps to retain the fine gold through amalgamation. If the gold is not clean, it will not amalgamate.

It may be necessary to elevate parts of the sluice on trestles or other devices to maintain approximately a grade of six inches drop for each twelve feet of sluice.

The riffles should not be fastened in the sluice box permanently as it is necessary to remove them for the clean up. They may be held in place by nailing the side boards of the box to the ends of the riffles. The nail should not be driven all the way in. Or they may be wedged in place.

---

* Longridge, C. C., Ibid. p. 194.

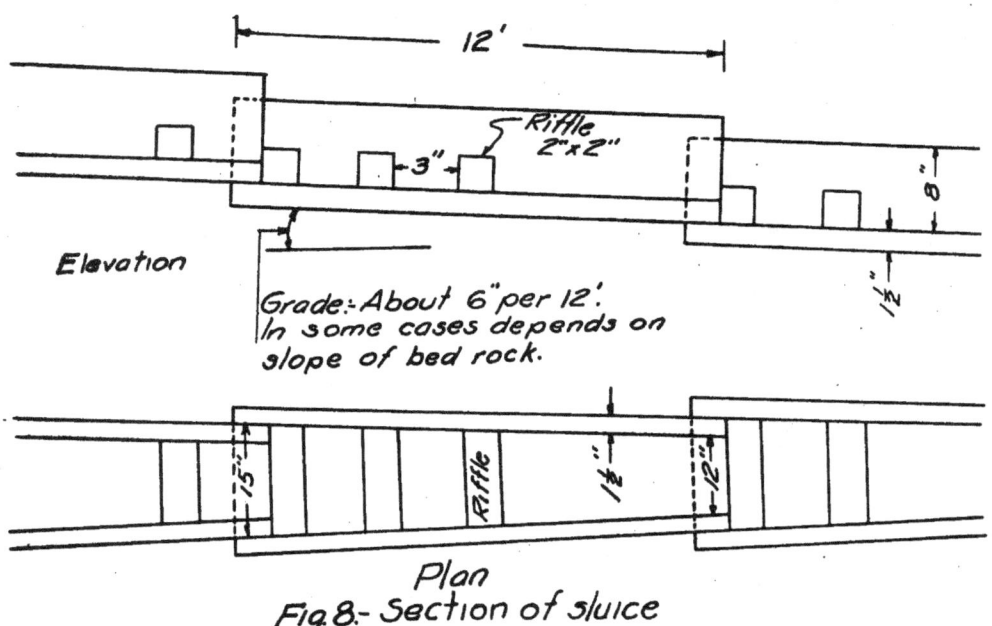

12'

Riffle 2"x2"

3"

8"

½"

Elevation

Grade:- About 6" per 12'.
In some cases depends on
slope of bed rock.

15"

Riffle

14"

12"

Plan
Fig. 8.- Section of sluice

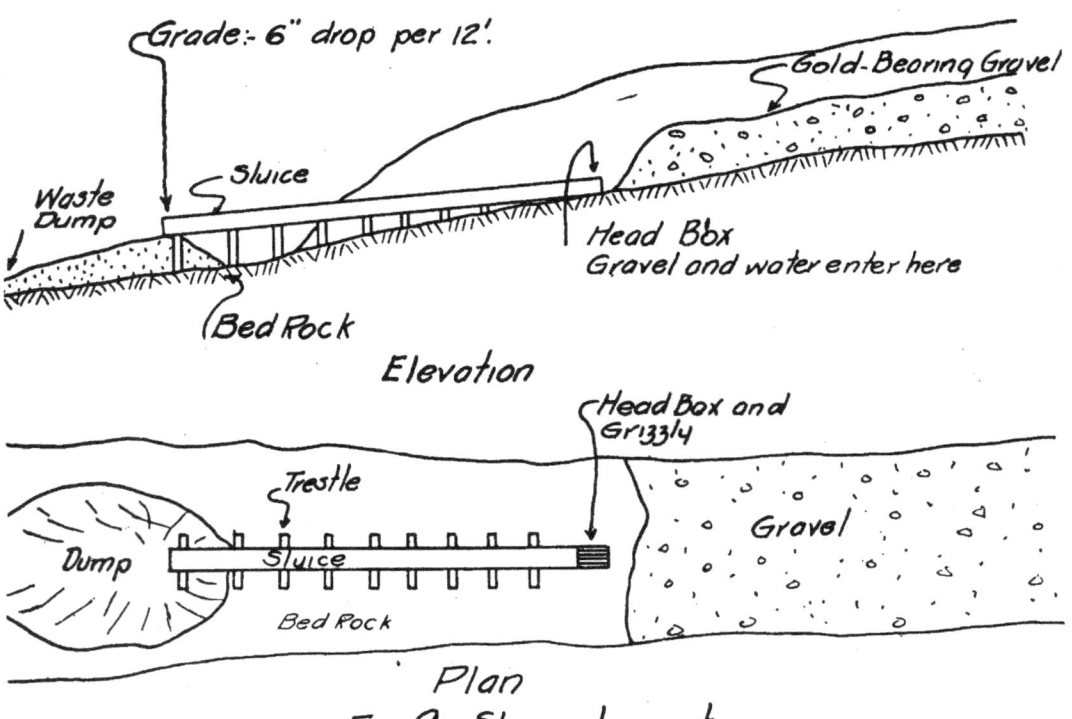

Grade:- 6" drop per 12'.

Gold-Bearing Gravel

Sluice

Waste
Dump

Head Box
Gravel and water enter here

Bed Rock

Elevation

Head Box and
Grizzly

Trestle

Dump

Sluice

Gravel

Bed Rock

Plan
Fig. 9.- Sluice lay-out.

— 14 —

Clean Up

The frequency of the clean up depends upon the richness of the gravel being washed. It may vary from a few days to the entire season. The first few riffles should be cleaned up at least once every two weeks. In making the clean up the gravel is discontinued and a stream of water, just large enough to wash the heavy sands, mercury, and amalgam, is permitted to flow down the sluice. The riffles are taken up and the sand washed down the sluice. Occasionally, the contents are scraped with a spoon. All cracks and crevices are thoroughly cleaned. Blankets and burlap that may have been used are washed in a tub.

Operation

In order to use a sluice, plenty of water must be available as a continuous stream is run through the system. If sufficient water is not at hand, it is useless to construct the sluice. For large scale operations, water may be brought to the gold-bearing deposits by means of a flume.

The gravel is shoveled onto the grizzly at the head box and the water run over it. The over-size is raked or shoveled off to one side. The amount of water flowing down the sluice should be just enough to wash the gravel, passing through the grizzly, over the riffles, and out the end of the sluice. For this reason, the grizzly bars should not be spaced too far apart. If so, the velocity of the water may have to be so great as to prevent the settling of the fine gold. When the wodden riffles become so worn that they no longer hold back the heavy sands, they should be replaced. This condition exists when the riffles become rounded or are worn thin.

Figure 9 illustrates the method of working a gravel bed where it is not necessary to elevate the material.

RECOVERY OF FINE GOLD*

Very fine gold is usually recovered in one of two ways or a combination of both. These methods are the use of undercurrents and gold-saving tables. The essential difference between the two is that the tables are usually covered with carpet, burlap, hides, matting, or some similar mateial, and quite often have a flatter grade than do the undercurrents proper. They are also much wider.

Descriptions of these two additions to the main sluice follow.

Undercurrents

The conditions existing in the operation of the main sluice do not permit the settling of the fine gold. This is because of the comparatively high velocity necessary to move the large quantity of gravel and sand, and to prevent them from lodging and building up back of the riffles. It is essential that everything larger than the very fine gravel (about ¼ inch in size and preferably nothing larger than coarse sand, be excluded from the undercurrent. This is accomplished by inserting a grizzly or perforated iron plate near the end of the sluice and above the trough leading to the undercurrent.

The undercurrent consists of a series of shallow wooden sluices. Their width is eight to ten times the width of the main sluice. This fulfills one of the main requirements of the undercurrent, **a large decrease in velocity of the water.** The length of the undercurrent is two to four times its width. For example, a main sluice

---

* Longridge, C. C., Ibid, pp. 264, 266.

Wilson, E. B., Hydraulic and Placer Mining, p. 145.

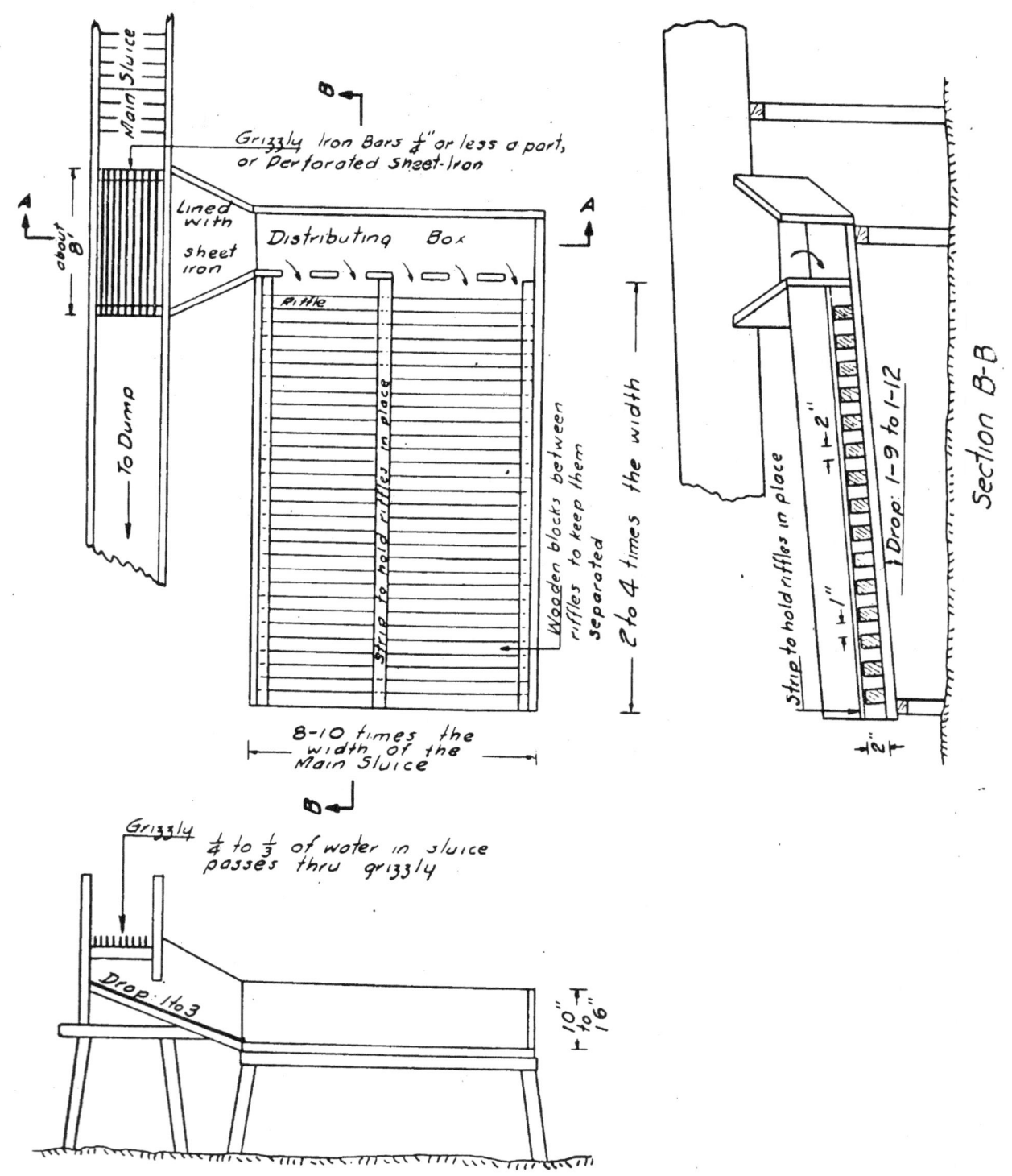

Main Sluice

Grizzly Iron Bars ¼" or less apart, or Perforated sheet-iron

about 8'

Lined with sheet iron

Distributing Box

To Dump

Riffle

Strip to hold riffles in place

Wooden blocks between riffles to keep them separated

2 to 4 times the width

8-10 times the width of the Main Sluice

Strip to hold riffles in place

1"

2"

Drop: 1-9 to 1-12

2"

Section B-B

Grizzly

¼ to ⅓ of water in sluice passes thru grizzly

Drop: 1 to 3

10" to 16"

Section A-A

Fig. 10- Undercurrent

12 inches wide should require an undercurrent of about 8 feet in width and about 20 feet long. The bottom of the undercurrent is made of planks about one and one-half inches thick. The joints must be tight. The sides are about 10 inches high. The bottom must be thickly covered with riffles. Material used for the riffles may be wooden strips, cobble stones, blocks, etc. They are spaced about one inch apart and are about two inches deep. The grade varies from one-foot drop in 12 feet of length to one-foot in nine feet. The exact grade depends on the type of riffle, size of gold, amount of water flowing, etc., and must be determined by experimenting with the conditions present. In some cases, the lower riffles of the undercurrent are replaced by an amalgamating plate.

It is very necessary that the sandy material flows over the undercurrent in a thin layer. Wide experience has shown that about ten per cent of the gold is recovered on undercurrents. In many instances, of course, it is much greater.

Figure 10 shows a sketch of the undercurrent.

Gold-Saving Tables

The construction of tables is identical with undercurrents with the exception of the material used for riffles. Burlap, carpet, blankets, hides, etc., are used. They are held in place by tacks and chicken wire, and, in some instances, by means of wooden strips. Only the fine sands should be permitted to pass over the gold tables, and they should do so in a thin film. The clean up is made by removing the covering and washing in a tub. At the end of the season the covering should be burned and the ashes panned for gold.

If wooden blocks are used on the undercurrent, they should be burned at the end of the season.

RECOVERY OF GOLD FROM SANDS‡

As the gold dust is mixed with more or less sand, iron, and other materials, it is necessary that it be cleaned. The larger pieces of foreign material are picked out by hand; the iron and magnetite are removed with a magnet. The finer sand can be removed by blowing it away. However, if this is done, there is danger of loosing the very fine gold.

If mercury has been used, the amalgam formed is softened with an excess of mercury and the mixture stirred. This procedure causes the base material to rise to the top where it can be skimmed off. The excess mercury is removed from the cleaned amalgam by squeezing through a chamois skin or strong, cotton cloth.

Cleaning Heavy Sands

The heavy material from the sluices, and from cleaning the gold dust and the amalgam, may contain other metals or minerals besides gold and amalgam. The most important of these are native copper, silver, platinum, iridosmine, monazite, pyrite, marcasite, hematite, chromite, galena, cinnabar, cassiterite, wolframite, scheelite, barite, and stibnite. Of the rock-forming minerals, the following may be present: Magnetite, ilmenite, rutile, garnet, zircon, tourmaline, and other.

As platinum does not amalgamate with the mercury, it will be left behind in the sands when the gold is amalgamated. The sands should, therefore, be carefully examined for flakes of platinum.

---

‡ Wimmler, N. L., Placer Mining Methods and Costs in Alaska; U. S. Bureau of Mines Bull. 259 (1927), p. 125.

When the fine gold is rusty or coated with materials which prevent is from amalgamating, it may sometimes be cleaned by agitating with a solution of cyanide and lye in a clean-up barrel.* This operation takes from 20 minutes to several hours, and then may not prove effective. The gold is brightened up by this procedure. The mercury may be added in the barrel at the same time.

Use of Cyanide†

If the cyandide is used too carelessly, solution of the gold will result. Solutions of certain strengths dissolve the gold more readily than others.

Maclaurin§ has found that the greatest amount of gold is dissolved in a solution of potassium cyanide of 0.25 per cent strength. A safe means of using cyanide is to make up a colution of one ounce of 98 per cent potassium cyanide to one-half gallon of water, and then use four ounces, or about one-half teacup, of this solution to 10 gallons of water.††

Retorting the Amalgam

If a retort is available, the cleaned amalgam is broken and packed loosely into the retort, which should have the inside coated with clay, chalk, or paper. The retort should not be more than three-quarters full. The cover must be fitted on tightly and sealed with either an asbestos gasket or with clay. The heating of the retort must progress slowly, the volatilization of the mercury not starting for about an hour. The iron pipe leading from the top of the retort must be kept cool by wrapping it in wet sacks. Water must continually be poured on the sacks. A dark red heat is about the proper temperture; at the end of the progress the temperature should be raised to a cherry red. The condenser pipe should not be put into a vessel of water. If this were done, and should the fire die down, the water would rush into the retort and cause a dangerous explosion. The retort must be allowed to cool gradually before opening. The outlet of the retort should be out of doors as the mercury fumes are very poisonous.

The small balls of amalgam obtained by the prospector are usually placed on a shovel and held over the fire to drive off the mercury. This should be done out of doors, and care should be taken that one does not breathe the fumes.

USE OF MERCURY IN PLACER MINING

Mercury may be used at various points in placer operations.
1. Back of the riffles in the main sluice.
2. In grooves or back of riffles on the undercurrent.
3. On the amalgamation plate at the discharge end of the undercurrent, or amalgamation plate in the sluice when only relatively f i n e material is passed through the boxes.
4. In the clean-up of the sluice-line.
5. In barrel amalgamation for dirty gold.
6. In the gold pan, either as liquid mercury or mercury-coated copper bottom.
Most of these applications may be in use at the same time.
Items 1, 2 and 6 are self-explanatory.
The following procedure may be followed for preparing t h e amalgamation plate.**

---

† Thomson, F. A., Stamp Milling and Cyaniding, 1st Ed. (1915), Chapters 8 and 10.
§ Maclaurin, J., The Dissolution of Gold in a Solution of Potassium Cyanide; Jour. Chem. Soc. (London), vol. 63, 1893, pp. 724-738; vol. 67, 1985, p. 199.
†† Wimmler, N. L., Ibid, p. 217.
* See Page 21.
** Vary, R. A., Amalgamation Practice at Porcupine United Gold Mines, Ltd., Timmins, Ont.; U. S. Bureau of Mines I. C. 6433 (March, 1931)

Amalgamation Plates**

The preparation of the amalgamation plates is done in the following steps:

1. Copper plate is thoroughly scrubbed with a solution of sodium hydroxide or lye to remove all signs of grease.

2. Wash the plate in clear water.

3. Thoroughly wash the plate with a dilute solution (about one ounce to one gallon of water) of sodium cyanide or potassium cyanide. This treatment should be continued until the copper surface is clean and bright.

4. Rub mercury on the plate with a whisk broom. When this is finished, there should be no copper showing, nor should the mercury be present in such excess that it appears in small wavelets or pools. The surface should appear moist and not dry and hard.

5. The mercury surface should have occasional treatment with the cyanide solution and fresh mercury should be added. Mercury amalgamates best with gold if there is already present a small amount of this metal. It is desirable, therefore, that a small amount of clean gold be added to mercury which has not as yet been used for amalgamating purposes.

6. Mercury should be shaken occasionally on the top of the plate during operations if the surface shows signs of becoming dry and hard.

Cleaning Amalgan from Plate*

1. Remove all particles of sand by sluicing down with clear water.

2. Brush the plate well with a stiff whisk broom, working from the bottom of the plate toward the top. If the surface is dry, mercury should be rubbed on before this is done.

3. Amalgam and mercury are taken from the top of the plate. The excess mercury is squeezed out through a heavy cotton cloth, or chamois skin, and the hard amalgam is retorted.

4. If the plate is too dry after the clean-up, mercury is shaken on and rubbed in. Then; starting at the bottom and working from the center toward the sides, the excess mercury is brushed to the top of the plate.

5. Washing with the dilute cyanide solution may be necessary to brighten up the surface after the clean-up.

6. In making the clean-up, care must be taken not to rub the plates too clean.

7. It is well to have a mercury trap (a deep, narrow trough) at the bottom of the plate to catch mercury and amalgam which break loose from the surface.

Sluice

In cleaning up the sluice, mercury may be used in the tub or receptacle in which the concentrates are caught. The wet material is thoroughly mixed and stirred with the mercury. This also applies to the use of mercury in the plain iron gold pan.

** Idem.

* Vary, R. A., Ibid.

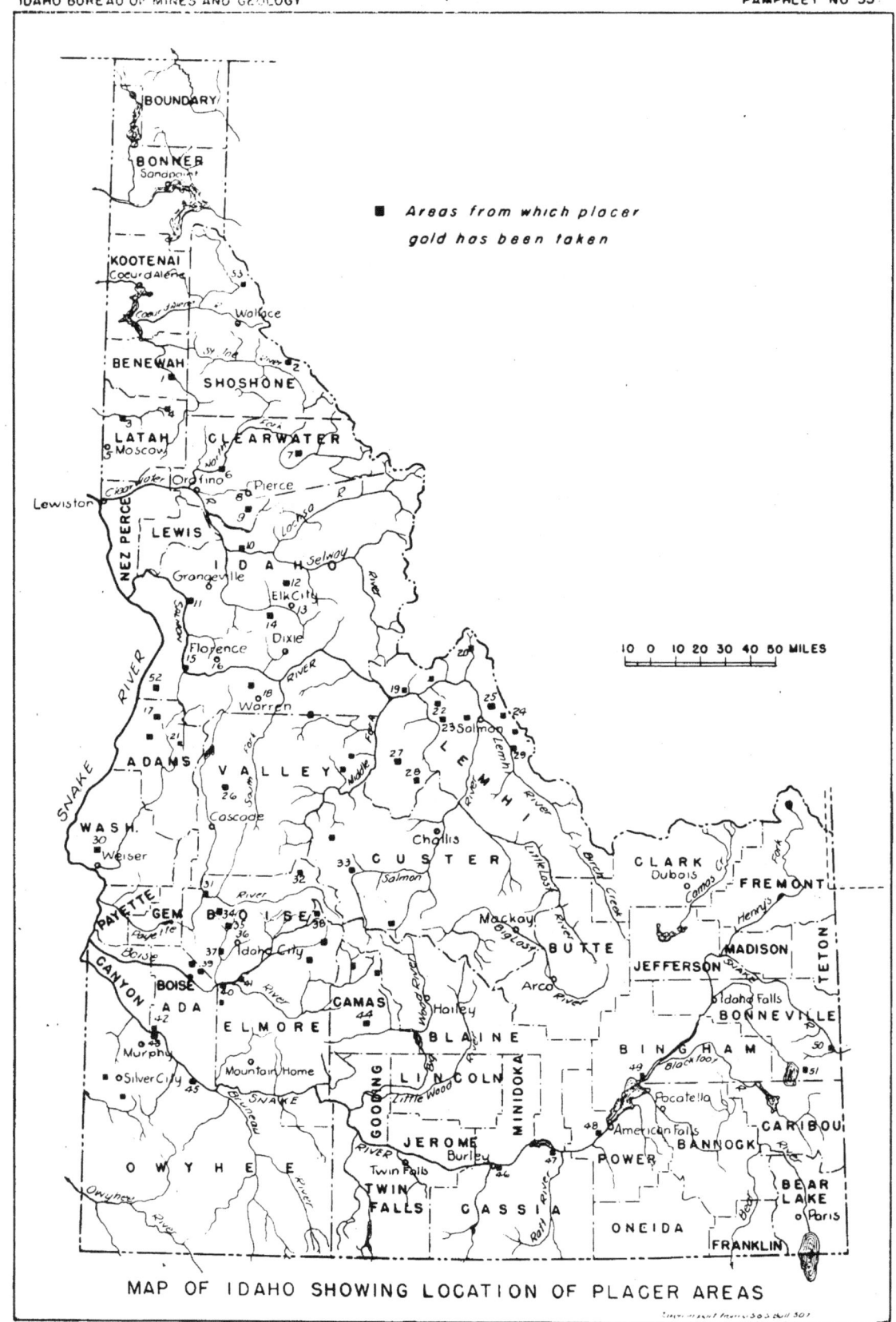

MAP OF IDAHO SHOWING LOCATION OF PLACER AREAS

Clean-up Barrel

The clean-up barrel is necessary when the gold is dirty and does not amalgamate easily. Concentrates, mercury, and weak cyanide solution† are placed in the barrel with a number of large (about 3-4 inches in diameter), clean rocks. The purpose of the rock is to polish the surface of the gold. The barrel is slowly rotated for an hour or more, depending on the condition of the gold. The cyanide solution must be very weak, otherwise the gold will dissolve and be lost. Sufficient mercury should be added to prevent the formation of a hard amalgam. The amount depends upon the quantity of gold present, and is best determined by experimenting.

Recovery of Gold from Amalgam

If the amalgam from the sluice, plate, or other source, is pasty or hard, sufficient mercury should be added to soften it. Then place the amalgam in a chamois skin and squeeze out the excess mercury. The greater the pressure, the more mercury is separated. A certain amount of gold remains dissolved in the mercury. It can only be obtained by distilling off the mercury. The amlgam is placed in an iron retort, which is gradually raised to a red heat. The mercury distills out, leaving behind impure gold. A retort may be constructed from a piece of iron pipe which has been threaded and plugged at one end; the other end is fitted with a union and condenser pipe bent so that the cooled mercury will run out the end. Wet burlap or cloth is wrapped around the condenser pipe. The end of the pipe should not be put under water.

DRY PLACER EQUIPMENT

Machines for operating dry placer deposits, so far as is known, have not been very successful. If a high grade deposit is available, and the gold fairly coarse, a fair saving may be made. The greater part will be blown away or will pass through the screen into the waste discard. The California Division of Mines Quarterly for April, 1932, contains information upon dry placer machines.

PLACER MINING IN IDAHO

The accompanying map shows the localities in which placer gold has been found in Idaho. **No assertion or prediction is made in this paper that gold may still be found in these localities.** In the early days of prospecting, Idaho w a s quite thoroughly worked over. It is not impossible that some pockets or streams were overlooked, or that in the years that have passed the gold lost in early operations has been reconcentrated. For this reason, the above mentioned map is included as a guide for the use of the inexperienced prospector for whom this brief paper has been written.

LIST OF MINING DISTRICTS TO ACCOMPANY SKETCH MAP OF IDAHO*

On the accompanying map no attempt has been made to show all of the streams or towns. To have done so would have caused unnecessary congestion. So far as the writer was able to determine, the map is reasonably complete in indicating the areas of known production of placer gold. An erroneous conclusion should not be drawn concerning this map. The map is not included as advocating that gold at the present day will be found in the various areas shown. It may be of help to the prospector in so far as a search for gold in a known territory may prove more fruitful than where placer gold has never been found. This should not, however, prevent further prospecting of districts which in the past have proved unfavorable.

---

† See Page 18 for making cyanide solution.

* Hill, J. M., The Mining Districts of the Western United States; U. S. Geological Survey Bull. 507 (1912).

PLACER MINING DISTRICTS OF IDAHO

No.	County	Mining District
1	Kootenai	Camas Cove (Tyson)
2	Shoshone	St. Joe
53	Shoshone	Beaver (Coeur d'Alene)
3	Latah	Gold Creek (Potlatch)
4	Latah	Hoodoo (Blackbird)
5	Latah	Moscow
6	Clearwater	Burnt Creek
7	Clearwater	Moose Creek
8	Clearwater	Pierce
9	Clearwater	Musselshell Creek (Weippe)
10	Idaho	Maggie
11	Idaho	Salmon River Placers (Simpson)
12	Idaho	Newsome
13	Idaho	Elk City
14	Idaho	Orogrande
15	Idaho	Salmon River Placers (Simpson)
16	Idaho	Florence
18	Idaho	Warren
52	Idaho	Crooks Corral
17	Adams	Black Lake
21	Adams	Meadows
19	Lemhi	Mineral Hill (Shoup)
20	Lemhi	Gibbonsville
22	Lemhi	Mackinaw
23	Lemhi	Leeburg (Arnett Creek)
24	Lemhi	Kirtley Creek
25	Lemhi	Pratt Creek
27	Lemhi	Yellowjacket
28	Lemhi	Gravel Range (Forney)
29	Lemhi	McDevitt
26	Boise	Gold Fork (Roseberry)
31	Boise	Payette River Placers (Jacobs Gulch)
32	Boise	Deadwood
34	Boise	Quartzburg (Idaho Basin)
35	Boise	Centerville (Idaho Basin)
36	Boise	Idaho City (Idaho Basin)
37	Boise	Monroe Creek
41	Boise	Twin Springs
30	Washington	Monroe Creek (Weiser)
33	Custer	Stanley Basin
38	Elmore	Atlanta
40	Elmore	Highland Valley
39	Ada	Black Hornet (Highland Valley, Shaw Mountain)
42	Ada	Snake River Placers
43	Owyhee	Snake River Placers
45	Owyhee	Snake River Placers
44	Blaine	Soldier
48	Blaine	Snake River Placers
46	Cassia	Snake River Placers
47	Cassia	Snake River Placers
49	Bingham	Snake River Placers
50	Bonneville	Snake River Placers
51	Bonneville	Mt. Pisgah (Caribou)

APPENDIX

IDAHO STATE MINING LAWS RELATING TO PLACER DEPOSITS*

For the benefit of those who are not familiar with the State mining laws regarding placer locations, the reproduction of part of the law is given here. If greater detail is desired, the reader is advised to get a copy of the Mining Laws of the State of Idaho which may be obtained from the State Mine Inspector, Boise, Idaho.

Placer Claims

Paragraph 5535 (3221) Location of placer claims. Placer claims, as mentioned in section 2329 of the Revised Statutes of the United States, may be located for the purpose of mining deposits and precious stones after discovery of such deposits.

Paragraph 5536 (3222) Monuments: Notice: Excavation: Record of notice. The locator of any placer mining claim located for the purpose of mining placer deposits or precious stones must, at the time of making the location, place a substantial post or monument, as is required in the location of quartz claims, at each corner of the location, and must also post at one of the same a notice of location containing the date of the location, the name of the locator, the name and dimensions of the claim, the mining district (if any) and county in which the same is situated; and must also give the distance and direction from said post or monument to such natural object or permanent monument, if any such there be, as will fix and describe in the notice itself the location of the claim. Within 15 days after making the location, the locator must make an excavation upon the claim of not less than 100 cubic feet, for the purpose of prospecting the same. Within 30 days after the location, the locator must file for record in the office of the recorder of the county, or the deputy recorder of the mining district in which the claim is situated, a substantial copy of his copy of notice of location, to which must be attached an affidavit such as is required in case of quartz claims.

Extracts from United States Code Compact Edition
(Title 30, Chapter 2)

Paragraph 35. Placer claims conforming entry to legal subdivisions and surveys: Limitations of claims. Claims usually called "placers," including all forms of deposit, excepting veins of quartz, or other rock in place, shall be subject to entry and patent, under like curcumstances and conditions, and upon similar proceedings, as are provided for vein or lode claims, but where the lands have been previously surveyed by the United States, the entry in its exterior limits shall conform to the legal subdivisions of the public lands. And where placers are upon surveyed lands, and conform to legal subdivisions, no further survey or plat shall be required, and all placer-mining claims located after the 10th day of May, 1872, shall conform as near as practicable with the United States system of public-land surveys, and the rectangular subdivisions of such surveys, and no such location shall include more than 20 acres for each individual claimant, but where placer claims can not be conformed to legal subdivisions, survey and plat shall be made as on unsurveyed lands; and where by the segregation of mineral land in any legal subdivision a quantity of agricultural land less than 40 acres remains, such fractional portion of agricultural land may be entered by any party qualified by law, for homestead purposes.

Paragraph 36. Same: Subdivisions of 10-acre tracts; maximum placer locations. Legal subdivision of 40 acres may be subdivided into 10-acre tracts; and two or more persons, or associations of persons, having contiguous claims of any size, although such claims may be of less than 10 acres each, may make joint entry thereof; but no location of a placer claim, made after the 9th day of July, 1870, shall exceed 160 acres for any one person or association of persons, which location shall conform to the United States surveys; and nothing in this section contained shall defeat or impair any bona fide preemption or homestead claim upon agricultural lands, or authorize the sale of the improvements of any bona fide settler to any purchaser.

IDENTIFICATION OF MINERALS
COMMONLY OCCURRING WITH GOLD IN PLACER DEPOSITS *

For the benefit of those who are not familiar with the minerals listed on the following pages of this report, the ensuing information is presented.

Amalgam

An alloy of gold and quicksilver and frequently silver. May contain copper. Color, silver white. Usually liquid but may be solid if there is an excess of gold and silver.

Barite

Heavy spar. Barytes. (Barium sulfate). Brittle. Hardness equals 2.5-3.5. Specific gravity equals 4.3-4.6. Color, white; also may be yellow, gray, blue, red, brown, or dark brown. Transparent to opaque. Characterized by high specific gravity, insolubility in acids, and cleavage.

Cassiterite

Tin stone. Stream tin. Tin ore (tin dioxide). Brittle. Hardness equals 6-7. Specific gravity equals 6.8-7.1. Color, brown or black, sometimes red, gray, white or yellow. Distinguished because of high gravity, hardness, and infusibility.

Chromite

(Iron oxide and chromium oxide.) Brittle. Hardness equals 5.5. Specific gravity equals 4.3-4.6. Has a metallic luster. Color, between iron-black and brownish-black. Sometimes feebly magnetic. Insoluble in acids.

Cinnebar

(Mercury sulfide.) Hardness equals 2-2.5. Specific gravity equals 8. Has a metallic luster. Color, cochineal-red, brownish-red, a n d lead-gray. Powder has scarlet color. Characterized by its color and high specific gravity, and softness.

Copper

Very ductile and malleable. Hardness equals 2.5-3. Specific gravity equals 8.8. Has a metallic luster. Color, copper-red.

Galena

Galenite. Lead glance (lead sulfide). Usually occurs in cubes. Hardness equals 2.5. Specific gravity equals 7.5. Has metallic luster. Color, lead-gray. Distinguished by color, softness, high specific gravity, and usually cubic cleavage.

Garnet

(Silicates that may contain calcium, magnesium, iron, aluminum, manganese, chromium, or titanium). Usually occurs in crystal-line form. The variety grossularite may be massive without apparent crystal form. Brittle to tough when massive. Hardness equals 6.5-7.5. Specific gravity equals 3.1-4.3. Has a resinous luster. Color, red, brown, yellow, white, apple-green, black; some bright red and green colors; white, when finely powdered.

---

* Ford, W. E., Dana's Textbook of Mineralogy. 3rd Ed. (1922).

Gold

Very malleable and ductile. Hardness equals 2.5-3. Specific gravity equals 15.6-19.3. When pure, equals 19.3. Has a metallic luster. Color, gold-yellow, sometimes silver-white; rarely orange-red. Usually alloyed with silver in varying amounts. Distinguished from pyrite and mica by softness and malleability, high specific gravity, and insolubility in acids. Chalcopyrite and pyrite may be confused with gold. They are both brittle and soluble in nitric acid. Usually occurs in placer deposits as flattened scales.

Hematite

(Iron oxide.) Specular hematitie would be the variety most likely to be found in placers. Brittle. Laminated flaky structure. Hardness equals 5.5-6.5. Specific gravity equals 4.9-5.3. Has a metallic luster. Streak has cherry-red or reddish brown color. Color, dark steel-gray or iron-black, or red. When sample is scraped with a knife, small, black, sparkling flakes drop.

Ilmenite

Menaccanite. Titanic iron ore. (Iron titanium oxide.) Occurs in placer as grains. Hardness equals 5-6. Specific gravity equals 4.5-5. Has a somewhat metallic luster. Streak is black to brownish red in color. Color, iron-black. Very slightly magnetic.

Magnetite

Magnetic iron ore. (Iron oxide.) Brittle. Hardness equals 5.5-6.5. Specific gravity equals 5. Has metallic luster. Streak, black. Very strongly magnetic. Sometimes is a magnet itself. Distinguished by being readily attracted by a magnet.

Marcasite

White iron pyrite. (Iron sulphide.) Brittle. Hardness equals 6-6.5. Specific gravity equals 4.9. Has metallic luster. Color, pale bronze-yellow. Streak, grayish or brownish black. Has lighter color than pyrite.

Monazite

(Cerium, lanthanum, thorium phosphate.) Usually occurs in grains. Sometimes flattened. Brittle. Hardness equals 5-5.5. Specific gravity equals 4.9-5.3. Has a resinous luster. Color, hyacinth-red, clove-brown, reddish or yellowish brown. Slightly transparent.

Platinum

(Alloyed with iron, iridium, rhodium, palladium, etc.) Usually in grains or scales. Malleable and ductile. Hardness equals 4-4.5. Specific gravity equals 14-19. When pure, 21-22. Has a metallic luster. Color, whitish steel-gray; shiney. Occasionally magnetic (if high in iron). Distinguish by color, high gravity, malleability, and insolubility in acids.

Pyrite

Iron pyrite. (Iron sulphide.) Brittle. Hardness equals 6-6.5. Specific gravity equals 4.9-5.1. Has metallic, glistening luster. Color, a pale brass-yellow. Streak, greenish black or brownish black. Quite often occurs as cubes.

Rutile

(Titanium dioxide.) Brittle. Hardness equals 6-6.5. Specific gravity equals 4.25. Has metallic luster. Color, reddish-brown to red; sometimes yellowish, bluish, violet, black. Powder, pale brown.

Scheelite

(Calcium tungstate.) Brittle. Hardness equals 4.5-5. Specific gravity equals 5.9-6.1. Color, white, yellowish-white, pale yellow, brownish, greenish, reddish. Powder, white.

Silver

Ductile and malleable. Hardness equals 2.5-3. Specific gravity equals 10.1-11.1. Pure, 10.5. Has metallic luster. Color, silver white; sometimes gray to black from tarnish. May contain some gold, copper, antimony, bismuth, or mermucy.

Stibnite

Antimonite, antimony glance. (Antimony trisulphide.) Hardness equals 2. Specific gravity equals 4.5. Metallic luster, sparkling appearance on fresh surface. Color, lead-gray. Streak, lead-gray.

Tourmaline

(Boron and aluminum silicate.) Brittle. Hardness equals 7-7.5. Specific gravity equals 2.9-3.2. Luster, vitreous to resinous. Color, black, brownish-black, bluish-black; may be blue, green, red, white, or colorless. Usually has a triangular-looking cross section.

Wolframite

(Iron, manganese tungstate.) Brittle. Hardness equals 5-5.5. Specific equals 7.2-7.5. Luster, sub-metallic. Color, dark grayish or brownish-black. Streak, nearly black. Sometimes weakly magnetic.

Zircon

(Zirconium silicate.) Brittle. Hardness equals 7.5. Specific gravity equals 4.7. Color, colorless, pale yellowish, grayish, yellowish-green, brownish-yellow, reddish-brown. Streak, uncolored.

EXPLANATION OF TERMS

The relative hardness of a mineral can be determined as follows:

The finger nail scratches minerals with a hardness of 2.

Those with a hardness of 3 are easily cut with a knife.

Minerals with a hardness of 4 are rather easily scratched with a knife.

Those minerals with a hardness of 5 are scratched with difficulty by a knife.

Hardness of 6 is barely scratched with a knife, but easily with a file. These minerals scratch glass.

Minerals with a hardness of 7 (for example, quartz) or over, scratch easily, but are barely scratched with a file.

In determining hardness, use is made of the following:

The finger nail was a hardness of 2.

A copper cent has a hardness of about 3.

The ordinary pocket knife is just over 5.

Ordinary window glass has a hardness of 5.5.

A piece of an unglazed dish, plate, or cup, is suitable for determining streak.

Or the mineral may be finely powdered.

BIBLIOGRAPHY

Lock, A. G., Gold: Its Occurrence and Extraction. (1882) Published by E. and F. N. Spon, London and New York.

*Wilson, E. B., Hydraulic and Placer Mining. 3rd Edition (1918) Published by John Wiley & Sons, Inc., New York.

*   The Gold Placers of Parts of Seward Peninsula, Alaska. (1908) U. S. Geological Survey Bulletin 328.

*Longridge, C. C., Hydraultic Mining. (1910) The Mining Journal, London.

*Brooks, A. H., Reconnaissance in the Cape Nome and Norton Bay Region, Alaska. Special Pub., U. S. Geological Survey, p. 146. (1901)

Young, G. J., Elements of Mining. 2nd Edition. (1923) Published by the McGraw-Hill Book Company, New York.

Peele, R., Mining Engineer's Handbook. 1st Edition. (1918) Vol. 1. Published by John Wiley & Sons, New York.

Storms, W. H., How to Make a Rocker. Eng. & Min. Jour., June 24, 1911, p. 1243.

Wimmler, N. L., Placer Mining Methods and Costs in Alaska. U. S. Bureau of Mines Bulletin 259, p. 215. (1927)

*Thomson, F. A., Stamp Milling and Cyaniding. 1st Edition. (1915) Chapters 8 and 10. Published by McGraw-Hill Book Company, New York.

*Maclaurin, J., The Dissolution of Gold in a Solution of Potassium Cyanide. Jour. Chem. Soc., (London) Vol. 63 (1893), pp. 724-738; Vol. 67 (1895), p. 199.

Janin, C., Placer Mining Methods and Operating Costs. U. S. Bureau of Mines Bulletin 121 (1916)

*Purington, C. W., Methods and Costs of Gravel and Placer Mining in Alaska. U. S. Geological Survey Bulletin 263. (1905).

Gardner, W. H., Drilling for Placer Gold. Published by Keystone Driller Company, Beaver Falls, Pa.

Ransome, F. L., Geology and Ore Deposits of the Breckenridge District, Colorado. U. S. Geological Survey Professional Paper 75. (1911)

Knox, H. B., and Haley, C. S., The Mining of Alluvial Deposits. The Mining Journal (London), Vol. 12, No. 2, p. 89; Vol. 12, No. 3, p. 153; Vol. 12, No. 4, p. 211. (1915)

Ellis, H. L., Prospecting Methods at Fairbanks. Eng. & Min. Jour., vol. 99, No. 19, p. 805 (May 8, 1915)
*   Annual Report of the State Inspector of Mines on the Mining Industry of Idaho.

Ford, W. E., Dana's Textbook of Mineralogy. 3rd Edition (1922)
        Mining Laws of the State of Idaho (May 8, 1929).

Hill, J. M., The Mining Districts of Western United States; U. S. Geological Survey Bulletin 507 (1912)

Hayes, C. W., and Lindgren, W., Contributions to Economic Geology; U. S. Geological Survey Bulletin 470 (1910)

Ransome, F. L., and Gale, H. S., Contributions to Economic Geology; U. S. Geological Survey Bulletin 580 (1913)

Ransome, F. L., and Gale, H. S., Contributions to Economic Geology; U. S. Geological Survey Bulletin 620 (1915)

* Publications especially interesting and instructive.

Umpleby, J. B., Geology and Ore Deposits of Lemhi County, Idaho; U. S. Geological Survey Bulletin 528 (1913)

Van Wagener, T. F., Manual of Hydraulic Mining for the Use of the Practical Miner. 3rd Edition Revised (1900). Published by D. Van Nostrand Company, New York.

Haley, C. S., Gold Placers of California; California State Mining Bureau Bulletin 92 (1923).

Vary, R. A., Amalgamation Practice at Porcupine United Gold Mines, Ltd., Timmins, Ontario; U. S. Bureau of Mines I. C. 6433 (1931)

Stolfa, L., Prospecting for Gold (Published by the author, Cicero, Ill.)

Boericke, W. S., Prospecting and Operating Small Gold Placers. Published by John Wiley & Sons, New York.

Sur, F. J., Placer Gold Mining. Published by Stanley Rose, Hollywood, Calif. Mining in California. Quarterly chapter of State Mineralogist's Report XXVIII, April, 1932. California Division of Mines, Ferry Building, San Francisco, California.

Wells, E. H., and Wooton, T. P., Gold Mining and Gold Deposits in New Mexico; Circular No. 5, New Mexico School of Mines, Socorro, New Mexico.

Wilson, E. D., and Tenney, J. B., Arizona Gold Placers and Placering; University of Arizona Bulletin No. 132, Arizona Bureau of Mines, Mineral Technology Series No. 34, Tucson, Arizona.

Dingman, O. A., Placer Mining Possibilities in Montana; Bureau of Mines and Geology Memoir No. 5, Butte, Montana.

Ingersol, G. E., Hand Methods of Placer Mining and Placer Districts of Washington and Oregon; Washington State College Engineering Bulletin No. 40, Pullman, Washington.

Ingersol, G. E., The W.S.C. Placer Mill; Mines Information Bureau Circular No. 2, Washington State College, Pullman, Washington.

The following publication issued by the Idaho Bureau of Mines and Geology contain some information concerning placer gold:

Umpleby, J. B., and Livingston, D. C., A Reconnaissance in South-Central Idaho; Bulletin No. 3, pp. 13-17 (1920)

Thomson, F. A., and Ballard, S. M., Geology and Gold Resources of North-Central Idaho; Bulletin No. 7 (1924)

Ballard, S. M., Geology and Gold Resources of Boise Basin, Boise County, Idaho; Bulletin No. 9, pp. 13, 14, 31, 32, 33, 89. (1924)

Kirkham, V. R. D., and Ellis, E. W., Geology and Ore Deposits of Boundary County, Idaho; Bulletin No. 10, pp. 46, 51, 73. (1926)

Piper, A. M., and Laney, F. B., Geology and Metalliferous Resources of the Region about Silver City, Idaho; Bulletin No. 11, p. 51. (1926)

Finch, John W., Prospecting for Gold Ores; Pamphlet No. 37. (1932)

Fahrenwald, A. W., Recovery of Gold from Its Ores; Pamphlet No. 37. (1932)

For chemicals, mineral collections, blow-pipe outfits, etc., The Denver Fire Clay Company, Denver, Colorado, and the C. M. Fassett Company, Spokane, Washington, are suggested.

GOLD RUSH BOOKS

OREGON, USA

www.GoldMiningBooks.com

Books On Mining

Visit: www.goldminingbooks.com to order your copies or ask your favorite book seller to offer them.

Mining Books by Kerby Jackson

Gold Dust: Stories From Oregon's Mining Years - Oregon mining historian and prospector, Kerby Jackson, brings you a treasure trove of seventeen stories on Southern Oregon's rich history of gold prospecting, the prospectors and their discoveries, and the breathtaking areas they settled in and made homes. 5" X 8", 98 ppgs. Retail Price: $11.99

The Golden Trail: More Stories From Oregon's Mining Years - In his follow-up to "Gold Dust: Stories of Oregon's Mining Years", this time around, Jackson brings us twelve tales from Oregon's Gold Rush, including the story about the first gold strike on Canyon Creek in Grant County, about the old timers who found gold by the pail full at the Victor Mine near Galice, how Iradel Bray discovered a rich ledge of gold on the Coquille River during the height of the Rogue River War, a tale of two elderly miners on the hunt for a lost mine in the Cascade Mountains, details about the discovery of the famous Armstrong Nugget and others. 5" X 8", 70 ppgs. Retail Price: $10.99

Oregon Mining Books

Geology and Mineral Resources of Josephine County, Oregon - Unavailable since the 1970's, this important publication was originally compiled by the Oregon Department of Geology and Mineral Industries and includes important details on the economic geology and mineral resources of this important mining area in South Western Oregon. Included are notes on the history, geology and development of important mines, as well as insights into the mining of gold, copper, nickel, limestone, chromium and other minerals found in large quantities in Josephine County, Oregon. 8.5" X 11", 54 ppgs. Retail Price: $9.99

Mines and Prospects of the Mount Reuben Mining District - Unavailable since 1947, this important publication was originally compiled by geologist Elton Youngberg of the Oregon Department of Geology and Mineral Industries and includes detailed descriptions, histories and the geology of the Mount Reuben Mining District in Josephine County, Oregon. Included are notes on the history, geology, development and assay statistics, as well as underground maps of all the major mines and prospects in the vicinity of this much neglected mining district. 8.5" X 11", 48 ppgs. Retail Price: $9.99

The Granite Mining District - Notes on the history, geology and development of important mines in the well known Granite Mining District which is located in Grant County, Oregon. Some of the mines discussed include the Ajax, Blue Ribbon, Buffalo, Continental, Cougar-Independence, Magnolia, New York, Standard and the Tillicum. Also included are many rare maps pertaining to the mines in the area. 8.5" X 11", 48 ppgs. Retail Price: $9.99

Ore Deposits of the Takilma and Waldo Mining Districts of Josephine County, Oregon - The Waldo and Takilma mining districts are most notable for the fact that the earliest large scale mining of placer gold and copper in Oregon took place in these two areas. Included are details about some of the earliest large gold mines in the state such as the Llano de Oro, High Gravel, Cameron, Platerica, Deep Gravel and others, as well as copper mines such as the famous Queen of Bronze mine, the Waldo, Lily and Cowboy mines. This volume also includes six maps and 20 original illustrations. 8.5" X 11", 74 ppgs. Retail Price: $9.99

Metal Mines of Douglas, Coos and Curry Counties, Oregon - Oregon mining historian Kerby Jackson introduces us to a classic work on Oregon's mining history in this important re-issue of Bulletin 14C Volume 1, otherwise known as the Douglas, Coos & Curry Counties, Oregon Metal Mines Handbook. Unavailable since 1940, this important publication was originally compiled by the Oregon Department of Geology and Mineral Industries includes detailed descriptions, histories and the geology of over 250 metallic mineral mines and prospects in this rugged area of South West Oregon. 8.5" X 11", 158 ppgs. Retail Price: $19.99

Metal Mines of Jackson County, Oregon - Unavailable since 1943, this important publication was originally compiled by the Oregon Department of Geology and Mineral Industries includes detailed descriptions, histories and the geology of over 450 metallic mineral mines and prospects in Jackson County, Oregon. Included are such famous gold mining areas as Gold Hill, Jacksonville, Sterling and the Upper Applegate. 8.5″ X 11″, 220 ppgs. Retail Price: $24.99

Metal Mines of Josephine County, Oregon - Oregon mining historian Kerby Jackson introduces us to a classic work on Oregon's mining history in this important re-issue of Bulletin 14C, otherwise known as the Josephine County, Oregon Metal Mines Handbook. Unavailable since 1952, this important publication was originally compiled by the Oregon Department of Geology and Mineral Industries includes detailed descriptions, histories and the geology of over 500 metallic mineral mines and prospects in Josephine County, Oregon. 8.5″ X 11″, 250 ppgs. Retail Price: $24.99

Metal Mines of North East Oregon - Oregon mining historian Kerby Jackson introduces us to a classic work on Oregon's mining history in this important re-issue of Bulletin 14A and 14B, otherwise known as the North East Oregon Metal Mines Handbook. Unavailable since 1941, this important publication was originally compiled by the Oregon Department of Geology and Mineral Industries and includes detailed descriptions, histories and the geology of over 750 metallic mineral mines and prospects in North Eastern Oregon. 8.5″ X 11″, 310 ppgs. Retail Price: $29.99

Metal Mines of North West Oregon - Oregon mining historian Kerby Jackson introduces us to a classic work on Oregon's mining history in this important re-issue of Bulletin 14D, otherwise known as the North West Oregon Metal Mines Handbook. Unavailable since 1951, this important publication was originally compiled by the Oregon Department of Geology and Mineral Industries and includes detailed descriptions, histories and the geology of over 250 metallic mineral mines and prospects in North Western Oregon. 8.5″ X 11″, 182 ppgs. Retail Price: $19.99

Mines and Prospects of Oregon - Mining historian Kerby Jackson introduces us to a classic mining work by the Oregon Bureau of Mines in this important re-issue of The Handbook of Mines and Prospects of Oregon. Unavailable since 1916, this publication includes important insights into hundreds of gold, silver, copper, coal, limestone and other mines that operated in the State of Oregon around the turn of the 19th Century. Included are not only geological details on early mines throughout Oregon, but also insights into their history, production, locations and in some cases, also included are rare maps of their underground workings. 8.5″ X 11″, 314 ppgs. Retail Price: $24.99

Lode Gold of the Klamath Mountains of Northern California and South West Oregon
(See California Mining Books)

Mineral Resources of South West Oregon - Unavailable since 1914, this publication includes important insights into dozens of mines that once operated in South West Oregon, including the famous gold fields of Josephine and Jackson Counties, as well as the Coal Mines of Coos County. Included are not only geological details on early mines throughout South West Oregon, but also insights into their history, production and locations. 8.5″ X 11″, 154 ppgs. Retail Price: $11.99

Chromite Mining in The Klamath Mountains of California and Oregon
(See California Mining Books)

Southern Oregon Mineral Wealth - Unavailable since 1904, this rare publication provides a unique snapshot into the mines that were operating in the area at the time. Included are not only geological details on early mines throughout South West Oregon, but also insights into their history, production and locations. Some of the mining areas include Grave Creek, Greenback, Wolf Creek, Jump Off Joe Creek, Granite Hill, Galice, Mount Reuben, Gold Hill, Galls Creek, Kane Creek, Sardine Creek, Birdseye Creek, Evans Creek, Foots Creek, Jacksonville, Ashland, the Applegate River, Waldo, Kerby and the Illinois River, Althouse and Sucker Creek, as well as insights into local copper mining and other topics. 8.5″ X 11″, 64 ppgs. Retail Price: $8.99

Geology and Ore Deposits of the Takilma and Waldo Mining Districts - Unavailable since the 1933, this publication was originally compiled by the United States Geological Survey and includes details on gold and copper mining in the Takilma and Waldo Districts of Josephine County, Oregon. The Waldo and Takilma mining districts are most notable for the fact that the earliest large scale mining of placer gold and copper in Oregon took place in these two areas. Included in this report are details about some of the earliest large gold mines in the state such as the Llano de Oro, High Gravel, Cameron, Platerica, Deep Gravel and others, as well as copper mines such as the famous Queen of Bronze mine, the Waldo, Lily and Cowboy mines. In addition to geological examinations, insights are also provided into the production, day to day operations and early histories of these mines, as well as calculations of known mineral reserves in the area. This volume also includes six maps and 20 original illustrations. 8.5″ X 11″, 74 ppgs. Retail Price: $9.99

Gold Mines of Oregon - Oregon mining historian Kerby Jackson introduces us to a classic work on Oregon's mining history in this important re-issue of Bulletin 61, otherwise known as "Gold and Silver In Oregon". Unavailable since 1968, this important publication was originally compiled by geologists Howard C. Brooks and Len Ramp of the Oregon Department of Geology and Mineral Industries and includes detailed descriptions, histories and the geology of over 450 gold mines Oregon. Included are notes on the history, geology and gold production statistics of all the major mining areas in Oregon including the Klamath Mountains, the Blue Mountains and the North Cascades. While gold is where you find it, as every miner knows, the path to success is to prospect for gold where it was previously found. **8.5" X 11", 344 ppgs. Retail Price: $24.99**

Mines and Mineral Resources of Curry County Oregon - Originally published in 1916, this important publication on Oregon Mining has not been available for nearly a century. Included are rare insights into the history, production and locations of dozens of gold mines in Curry County, Oregon, as well as detailed information on important Oregon mining districts in that area such as those at Agness, Bald Face Creek, Mule Creek, Boulder Creek, China Diggings, Collier Creek, Elk River, Gold Beach, Rock Creek, Sixes River and elsewhere. Particular attention is especially paid to the famous beach gold deposits of this portion of the Oregon Coast. **8.5" X 11", 140 ppgs. Retail Price: $11.99**

Chromite Mining in South West Oregon - Originally published in 1961, this important publication on Oregon Mining has not been available for nearly a century. Included are rare insights into the history, production and locations of nearly 300 chromite mines in South Western Oregon. **8.5" X 11", 184 ppgs. Retail Price: $14.99**

Mineral Resources of Douglas County Oregon - Originally published in 1972, this important publication on Oregon Mining has not been available for nearly forty years. Included are rare insights into the geology, history, production and locations of numerous gold mines and other mining properties in Douglas County, Oregon. **8.5" X 11", 124 ppgs. Retail Price: $11.99**

Mineral Resources of Coos County Oregon - Originally published in 1972, this important publication on Oregon Mining has not been available for nearly forty years. Included are rare insights into the geology, history, production and locations of numerous gold mines and other mining properties in Coos County, Oregon. **8.5" X 11", 100 ppgs. Retail Price: $11.99**

Mineral Resources of Lane County Oregon - Originally published in 1938, this important publication on Oregon Mining has not been available for nearly seventy five years. Included are extremely rare insights into the geology and mines of Lane County, Oregon, in particular in the Bohemia, Blue River, Oakridge, Black Butte and Winberry Mining Districts. **8.5" X 11", 82 ppgs. Retail Price: $9.99**

Mineral Resources of the Upper Chetco River of Oregon: Including the Kalmiopsis Wilderness - Originally published in 1975, this important publication on Oregon Mining has not been available for nearly forty years. Withdrawn under the 1872 Mining Act since 1984, real insight into the minerals resources and mines of the Upper Chetco River has long been unavailable due to the remoteness of the area. Despite this, the decades of battle between property owners and environmental extremists over the last private mining inholding in the area has continued to pique the interest of those interested in mining and other forms of natural resource use. Gold mining began in the area in the 1850's and has a rich history in this geographic area, even if the facts surrounding it are little known. Included are twenty two rare photographs, as well as insights into the Becca and Morning Mine, the Emmly Mine (also known as Emily Camp), the Frazier Mine, the Golden Dream or Higgins Mine, Hustis Mine, Peck Mine and others. **8.5" X 11", 64 ppgs. Retail Price: $8.99**

Gold Dredging in Oregon - Originally published in 1939, this important publication on Oregon Mining has not been available for nearly seventy five years. Included are extremely rare insights into the history and day to day operations of the dragline and bucketline gold dredges that once worked the placer gold fields of South West and North East Oregon in decades gone by. Also included are details into the areas that were worked by gold dredges in Josephine, Jackson, Baker and Grant counties, as well as the economic factors that impacted this mining method. This volume also offers a unique look into the values of river bottom land in relation to both farming and mining, in how farm lands were mined, re-soiled and reclamated after the dredges worked them. Featured are hard to find maps of the gold dredge fields, as well as rare photographs from a bygone era. **8.5" X 11", 86 ppgs. Retail Price: $8.99**

Quick Silver Mining in Oregon - Originally published in 1963, this important publication on Oregon Mining has not been available for over fifty years. This publication includes details into the history and production of Elemental Mercury or Quicksilver in the State of Oregon. **8.5" X 11", 238 ppgs. Retail Price: $15.99**

Mines of the Greenhorn Mining District of Grant County Oregon - Originally published in 1948, this important publication on Oregon Mining has not been available for over sixty five years. In this publication are rare insights into the mines of the famous Greenhorn Mining District of Grant County, Oregon, especially the famous Morning Mine. Also included are details on the Tempest, Tiger, Bi-Metallic, Windsor, Psyche, Big Johnny, Snow Creek, Banzette and Paramount Mines, as well as prospects in the vicinities in the famous mining areas of Mormon Basin, Vinegar Basin and Desolation Creek. Included are hard to find mine maps and dozens of rare photographs from the bygone era of Grant County's rich mining history. **8.5" X 11", 72 ppgs. Retail Price: $9.99**

Geology of the Wallowa Mountains of Oregon: Part I (Volume 1) - Originally published in 1938, this important publication on Oregon Mining has not been available for nearly seventy five years. Included are details on the geology of this unique portion of North Eastern Oregon. This is the first part of a two book series on the area. Accompanying the text are rare photographs and historic maps. 8.5" X 11", **92 ppgs. Retail Price: $9.99**

Geology of the Wallowa Mountains of Oregon: Part II (Volume 2) - Originally published in 1938, this important publication on Oregon Mining has not been available for nearly seventy five years. Included are details on the geology of this unique portion of North Eastern Oregon. This is the first part of a two book series on the area. Accompanying the text are rare photographs and historic maps. 8.5" X 11", **94 ppgs. Retail Price: $9.99**

Field Identification of Minerals For Oregon Prospectors - Originally published in 1940, this important publication on Oregon Mining has not been available for nearly seventy five years. Included in this volume is an easy system for testing and identifying a wide range of minerals that might be found by prospectors, geologists and rockhounds in the State of Oregon, as well as in other locales. Topics include how to put together your own field testing kit and how to conduct rudimentary tests in the field. This volume is written in a clear and concise way to make it useful even for beginners. **8.5" X 11", 158 ppgs. Retail Price: $14.99**

The Bohemia Mining District of Oregon - Originally published in 1900, this important publication on Oregon Mining has not been available for over a century. Included in this volume are important insights into the famous Bohemia Mining District of Oregon, including the histories and locations of important gold mines in the area such as the Ophir Mine, Clarence, Acturas, Peek-a-boo, White Swan, Combination Mine, the Musick Mine, The California, White Ghost, The Mystery, Wall Street, Vesuvius, Story, Lizzie Bullock, Delta, Elsie Dora, Golden Slipper, Broadway, Champion Mine, Knott, Noonday, Helena, White Wings, Riverside and others. Also included are notes on the nearby Blue River Mining District. **8.5" X 11", 58 ppgs. Retail Price: $9.99**

The Gold Fields of Eastern Oregon - Unavailable since 1900, this publication was originally compiled by the Baker City Chamber of Commerce Offering important insights into the gold mining history of Eastern Oregon, "The Gold Fields of Eastern Oregon" sheds a rare light on many of the gold mines that were operating at the turn of the 19th Century in Baker County and Grant County in North Eastern Oregon. Some of the areas featured include the Cable Cove District, Baisely-Elhorn, Granite, Red Boy, Bonanza, Susanville, Sparta, Virtue, Vaughn, Sumpter, Burnt River, Rye Valley and other mining districts. Included is basic information on not only many gold mines that are well known to those interested in Eastern Oregon mining history, but also many mines and prospects which have been mostly lost to the passage of time. Accompanying are numerous rare photos **8.5" X 11", 78 ppgs. Retail Price: $10.99**

Gold Mining in Eastern Oregon - Originally published in 1938, this important publication on Oregon Mining has not been available for over a century. Included in this volume are important insights into the famous mining districts of Eastern Oregon during the late 1930's. Particular attention is given to those gold mines with milling and concentrating facilities in the Greenhorn, Red Boy, Alamo, Bonanza, Granite, Cable Cove, Cracker Creek, Virtue, Keating, Medical Springs, Sanger, Sparta, Chicken Creek, Mormon Basin, Connor Creek, Cornucopia and the Bull Run Mining Districts. Some of the mines featured include the Ben Harrison, North Pole-Columbia, Highland Maxwell, Baisley-Elkhorn, White Swan, Balm Creek, Twin Baby, Gem of Sparta, New Deal, Gleason, Gifford-Johnson, Cornucopia, Record, Bull Run, Orion and others. Of particular interest are the mill flow sheets and descriptions of milling operations of these mines. **8.5" X 11", 68 ppgs. Retail Price: $8.99**

The Gold Belt of the Blue Mountains of Oregon - Originally published in 1901, this important publication on Oregon Mining has not been available for over a century. Included in this volume are rare insights into the gold deposits of the Blue Mountains of North East Oregon, including the history of their early discovery and early production. Extensive details are offered on this important mining area's mineralogy and economic geology, as well as insights into nearby gold placers, silver deposits and copper deposits. Featured are the Elkhorn and Rock Creek mining districts, the Pocahontas district, Auburn and Minersville districts, Sumpter and Cracker Creek, Cable Cove, the Camp Carson district, Granite, Alamo, Greenhorn, Robinsonville, the Upper Burnt River Valley and Bonanza districts, Susanville, Quartzburg, Canyon Creek, Virtue, the Copper Butte district, the North Powder River, Sparta, Eagle Creek, Cornucopia, Pine Creek, Lower Powder River, the Upper Snake River Canyon, Rye Valley, Lower Burnt River Valley, Mormon Basin, the Malheur and Clarks Creek districts, Sutton Creek and others. Of particular interest are important details on numerous gold mines and prospects in these mining districts, including their locations, histories, geology and other important information, as well as information on silver, copper and fire opal deposits. **8.5" X 11", 250 ppgs. Retail Price: $24.99**

Mining in the Cascades Range of Oregon - Originally published in 1938, this important publication on Oregon Mining has not been available for over seventy five years. Included in this volume are rare insights into the gold mines and other types of metal mines in the Cascades Mountain Range of Oregon. Some of the important mining areas covered include the famous Bohemia Mining District, the North Santiam Mining District, Quartzville Mining District, Blue River Mining District, Fall Creek Mining District, Oakridge District, Zinc District, Buzzard-Al Sarena District, Grand Cove, Climax District and Barron Mining District. Of particular interest are important details on over 100 mines and prospects in these mining districts, including their locations, histories, geology and other important information. 8.5" X 11", 170 ppgs. Retail Price: $14.99

Beach Gold Placers of the Oregon Coast - Originally published in 1934, this important publication on Oregon Mining has not been available for over 80 years. Included in this volume are rare insights into the beach gold deposits of the State of Oregon, including their locations, occurance, composition and geology. Of particular interest is information on placer platinum in Oregon's rich beach deposits. Also included are the locations and other information on some famous Oregon beach mines, including the Pioneer, Eagle, Chickamin, Iowa and beach placer mines north of the mouth of the Rogue River. 8.5" X 11", 60 ppgs. Retail Price: $8.99

Mineralogical Composition of the Sands of the Oregon Coast: From Coos Bay to the Columbia - Published in 1945, he text features hard to find information on the composition of the gold bearing black sands of the South West Oregon Coast, offering a unique insight to prospectors in search of Oregon's legendary beach gold. 104 ppgs, $9.99

Manganese Mining in Oregon - First released in 1942 and now out of print, this special reprint edition of "Manganese in Oregon" was originally published by the Oregon Department of Geology and Mineral Industries. The text features hard to find information on the mining of Manganese in Oregon, including details and maps of Oregon manganese mines and prospects. 108 ppgs, 9.99

Medford Oregon As A Mining Center - Written in 1912, this hard to find publication includes valuable insights into the mining history of South West Oregon. This small book contains interesting information on the gold, copper and mining industry in Southern Oregon as it existed just prior to World War One, shedding light on some of the important mines in the area. Included are rare photographs and vintage advertising of the day. 80 ppgs, 9.99

Mineral Resources of Curry County Oregon - First released in 1977 and now out of print, this special reprint edition of "Geology, Mineral Resources and Rock Materials of Curry County, Oregon" was originally published in cooperation of Curry County, Oregon and the Oregon Department of Geology and Mineral Industries. The text features hard to find information on not only the mining of gold and other metals in Curry County, but also aggregate mining in the area. 102 ppgs, 11.99

Origin of the Gold Bearing Black Sands of the Coast of South West Oregon - First released in 1943 and now out of print, this special reprint edition of "The Origin of the Black Sands of the South West Oregon Coast" was originally published by the Oregon Department of Geology and Mineral Industries. The text features hard to find information on the origin of the gold bearing black sands of the South West Oregon Coast, offering a unique insight to prospectors in search of Oregon's legendary beach gold. 52 ppgs, 8.99

South West Oregon Mining - Leading mining historian Kerby Jackson introduces us to six classic small mining publications on the Gold Mining Industry in Southern Oregon. This small book consists of a compilation of USGS J.S. Diller's "Mines of the Riddles Quadrangle", "The Rogue River Valley Coal Fields" and "Mineral Resources of the Grants Pass Quadrangle", the Grants Pass Commercial Club's rare publication "Mining in Josephine County, Oregon" and the USGS publication "The Distribution of Placer Gold in the Sixes River, South West Oregon". Also included is F.W. Libbey's legendary article on the Southern Oregon Mining Industry, "Lest We Forget", which appeared in the publication of the Oregon State Department of Geology and Mineral Industries in the early 1960's. This compilation offers a unique perspective on mining in South West Oregon and includes considerable information on mines in Josephine, Jackson and Coos Counties. 142 ppgs, 14.99

Geology and Mineral Resources of the Gasquet Quadrangle of California-Oregon - First published in 1953, it has been unavailable for over a century and sheds important light on the geological features and mineral resources of this portion of Northern California and Southern Oregon. 80 ppgs, 9.99

Idaho Mining Books

Gold in Idaho - Unavailable since the 1940's, this publication was originally compiled by the Idaho Bureau of Mines and includes details on gold mining in Idaho. Included is not only raw data on gold production in Idaho, but also valuable insight into where gold may be found in Idaho, as well as practical information on the gold bearing rocks and other geological features that will assist those looking for placer and lode gold in the State of Idaho. This volume also includes thirteen gold maps that greatly enhance the practical usability of the information contained in this small book detailing where to find gold in Idaho. **8.5" X 11", 72 ppgs. Retail Price: $9.99**

Geology of the Couer D'Alene Mining District of Idaho - Unavailable since 1961, this publication was originally compiled by the Idaho Bureau of Mines and Geology and includes details on the mining of gold, silver and other minerals in the famous Coeur D'Alene Mining District in Northern Idaho. Included are details on the early history of the Coeur D'Alene Mining District, local tectonic settings, ore deposit features, information on the mineral belts of the Osburn Fault, as well as detailed information on the famous Bunker Hill Mine, the Dayrock Mine, Galena Mine, Lucky Friday Mine and the infamous Sunshine Mine. This volume also includes sixteen hard to find maps. **8.5" X 11", 70 ppgs. Retail Price: $9.99**

The Gold Camps and Silver Cities of Idaho - Originally published in 1963, this important publication on Idaho Mining has not been available for nearly fifty years. Included are rare insights into the history of Idaho's Gold Rush, as well as the mad craze for silver in the Idaho Panhandle. Documented in fine detail are the early mining excitements at Boise Basin, at South Boise, in the Owyhees, at Deadwood, Long Valley, Stanley Basin and Robinson Bar, at Atlanta, on the famous Boise River, Volcano, Little Smokey, Banner, Boise Ridge, Hailey, Leesburg, Lemhi, Pearl, at South Mountain, Shoup and Ulysses, Yellow Jacket and Loon Creek. The story follows with the appearance of Chinese miners at the new mining camps on the Snake River, Black Pine, Yankee Fork, Bay Horse, Clayton, Heath, Seven Devils, Gibbonsville, Vienna and Sawtooth City. Also included are special sections on the Idaho Lead and Silver mines of the late 1800's, as well as the mining discoveries of the early 1900's that paved the way for Idaho's modern mining and mineral industry. Lavishly illustrated with rare historic photos, this volume provides a one of a kind documentary into Idaho's mining history that is sure to be enjoyed by not only modern miners and prospectors who still scour the hills in search of nature's treasures, but also those enjoy history and tromping through overgrown ghost towns and long abandoned mining camps. **8.5" X 11", 186 ppgs. Retail Price: $14.99**

Ore Deposits and Mining in North Western Custer County Idaho - Unavailable since 1913, this important publication was originally published by the Us Department of the Interior and has been unavailable for a century. Included are fine details on the geology, geography, gold placers and gold and silver bearing quartz veins of the mining region of North West Custer County, Idaho. Of particular interest is a rare look at the mines and prospects of the region, including those such as the Ramshorn Mine, SkyLark, Riverview, Excelsior, Beardsley, Pacific, Hoosier, Silver Brick, Forest Rose and dozens of others in the Bay Horse Mining District. Also covered are the mines of the Yankee Fork District such as the Lucky Boy, Badger, Black, Enterprise, Charles Dickens, Morrison, Golden Sunbeam, Montana, Golden Gate and others, as well as those in the Loon Mining District. **8.5" X 11", 126 ppgs. Retail Price: $12.99**

Gold Rush To Idaho - Unavailable since 1963, this important publication was originally published by the Idaho Bureau of Mines and has been unavailable for 50 years. "Gold Rush To Idaho" revisits the earliest years of the discovery of gold in Idaho Territory and introduces us to the conditions that the pioneer gold seekers met when they blazed a trail through the wilderness of Idaho's mountains and discovered the precious yellow metal at Oro Fino and Pierce. Subsequent rushes followed at places like Elk City, Newsome, Clearwater Station, Florence, Warrens and elsewhere. Of particular interest is a rare look at the hardships that the first miners in Idaho met with during their day to day existences and their attempts to bring law and order to their mining camps. **8.5" X 11", 88 ppgs. Retail Price: $9.99**

The Geology and Mines of Northern Idaho and North Western Montana - Unavailable since 1909, this important publication was originally published by the Us Department of the Interior and has been unavailable for a century. Included are fine details on the geology and geography of the mining regions of Northern Idaho and North Western Montana. Of particular interest is a rare look at the mines and prospects of the region, including those in the Pine Creek Mining District, Lake Pend Oreille district, Troy Mining District, Sylvanite District, Cabinet Mining District, Prospect Mining District and the Missoula Valley. Some of the mines featured include the Iron Mountain, Silver Butte, Snowshoe, Grouse Mountain Mine and others. **8.5" X 11", 142 ppgs. Retail Price: $12.99**

Mining in the Alturas Quadrangle of Blaine County Idaho - Unavailable since 1922, this important publication was originally published by the Idaho Bureau of Mines and has been unavailable for ninety years. Topics include the geology, rock formations and the formation of ore deposits in this important mining area of Idaho. Of particular focus is information on the local geology, quartz veins and ore deposits of this portion of Idaho. Included are hard to find details, including the descriptions and locations of numerous gold and silver mines in the area including the Silver King, Pilgrim, Columbia, Lone Jack, Sunbeam, Pride of the West, Lucky Boy, Scotia, Atlanta, Beaver-Bidwell and others mines and prospects. **8.5" X 11", 56 ppgs. Retail Price: $8.99**

Mining in Lemhi County Idaho - Originally published in 1913, this important book on Idaho Mining has not been available to miners for over a century. Included are rare insights into hundreds of gold, silver, copper and other mines in this famous Idaho mining area. Details include the locations, geology, history, production and other facts of the mines of this region, not only gold and silver hardrock mines, but also gold placer mines, lead-silver deposits, copper mines, cobalt-nickel deposits, tungsten and tin mines . It is lavishly illustrated with hard to find photos of the period and rare mining maps. Some of the vicinities featured include the Nicholia Mining District, Spring Mountain District, Texas District, Blue Wing District, Junction District, McDevitt District, Pratt Creek, Eldorado District, Kirtley Creek, Carmen Creek, Gibbonsville, Indian Creek, Mineral Hill District, Mackinaw, Eureka District, Blackbird District, YellowJacket District, Gravel Range District, Junction District, Parker Mountain and other mining districts. 8.5" X 11", 226 ppgs. Retail Price: $19.99

Mining in Shoshone County Idaho - First published in 1923, it has been unavailable for over a century and sheds important light on the mining history of Shoshone County, Idaho. Some of the topics include the history of mining in Shoshone County, a look at the local geology and ore characteristics of lead-silver deposits, zinc deposits, copper, antimony, gold and other minerals. Also included are insights into the history, production, characteristics and locations of numerous mines in the area. 198 ppgs, 15.99

Utah Mining Books

Fluorite in Utah - Unavailable since 1954, this publication was originally compiled by the USGS, State of Utah and U.S. Atomic Energy Commission and details the mining of fluorspar, also known as fluorite in the State of Utah. Included are details on the geology and history of fluorspar (fluorite) mining in Utah, including details on where this unique gem mineral may be found in the State of Utah. 8.5" X 11", 60 ppgs. Retail Price: $8.99

The Gold Hill Mining District of Utah - First published in 1935, it has been unavailable since those days and sheds important light on the mines, history and geology of Utah's Gold Hill Mining District. Included are rare insights into this important mining area, including the locations, histories and details of numerous mines. This volume is well illustrated with geological diagrams, as well as hard to find maps of some of the most important mines in this district. 202 ppgs., 19.99

The Mines, Miners and Minerals of Utah - First published in 1896, it has been unavailable since those days and sheds important light on the early mines and miners of Pioneer Utah, as well as the minerals which they won from the earth by laborious hard physical labor and sheer determination. Included are rare insights into the early mining history of Utah, as well details on hundreds of gold, silver and copper mines. 376 ppgs., 24.99

California Mining Books

The Tertiary Gravels of the Sierra Nevada of California - Mining historian Kerby Jackson introduces us to a classic mining work by Waldemar Lindgren in this important re-issue of The Tertiary Gravels of the Sierra Nevada of California. Unavailable since 1911, this publication includes details on the gold bearing ancient river channels of the famous Sierra Nevada region of California. 8.5" X 11", 282 ppgs. Retail Price: $19.99

The Mother Lode Mining Region of California - Unavailable since 1900, this publication includes details on the gold mines of California's famous Mother Lode gold mining area. Included are details on the geology, history and important gold mines of the region, as well as insights into historic mining methods, mine timbering, mining machinery, mining bell signals and other details on how these mines operated. Also included are insights into the gold mines of the California Mother Lode that were in operation during the first sixty years of California's mining history. 8.5" X 11", 176 ppgs. Retail Price: $14.99

Lode Gold of the Klamath Mountains of Northern California and South West Oregon - Unavailable since 1971, this publication was originally compiled by Preston E. Hotz and includes details on the lode mining districts of Oregon and California's Klamath Mountains. Included are details on the geology, history and important lode mines of the French Gulch, Deadwood, Whiskeytown, Shasta, Redding, Muletown, South Fork, Old Diggings, Dog Creek (Delta), Bully Choop (Indian Creek), Harrison Gulch, Hayfork, Minersville, Trinity Center, Canyon Creek, East Fork, New River, Denny, Liberty (Black Bear), Cecilville, Callahan, Yreka, Fort Jones and Happy Camp mining districts in California, as well as the Ashland, Rogue River, Applegate, Illinois River, Takilma, Greenback, Galice, Silver Peak, Myrtle Creek and Mule Creek districts of South Western Oregon. Also included are insights into the mineralization and other characteristics of this important mining region. 8.5" X 11", 100 ppgs. Retail Price: $10.99

Mines and Mineral Resources of Shasta County, Siskiyou County, Trinity County: California - Unavailable since 1915, this publication was originally compiled by the California State Mining Bureau and includes details on the gold mines of this area of Northern California. Also included are insights into the mineralization and other characteristics of this important mining region, as well as the location of historic gold mines. 8.5" X 11", 204 ppgs. Retail Price: $19.99

Geology of the Yreka Quadrangle, Siskiyou County, California - Unavailable since 1977, this publication was originally compiled by Preston E. Hotz and includes details on the geology of the Yreka Quadrangle of Siskiyou County, California. Also included are insights into the mineralization and other characteristics of this important mining region. 8.5" X 11", 78 ppgs. Retail Price: $7.99

Mines of San Diego and Imperial Counties, California - Originally published in 1914, this important publication on California Mining has not been available for a century. This publication includes important information on the early gold mines of San Diego and Imperial County, which were some of the first gold fields mined in California by early Spanish and Mexican miners before the 49ers came on the scene. Included are not only details on early mining methods in the area, production statistics and geological information, but also the location of the early gold mines that helped make California "The Golden State". Also included are details on the mining of other minerals such as silver, lead, zinc, manganese, tungsten, vanadium, asbestos, barite, borax, cement, clay, dolomite, fluospar, gem stones, graphite, marble, salines, petroleum, stronium, talc and others. 8.5" X 11", 116 ppgs. Retail Price: $12.99

Mines of Sierra County, California - Unavailable since 1920, this publication was originally compiled by the California State Mining Bureau and includes details on the gold mines of Sierra County, California. Also included are insights into the mineralization and other characteristics of this important mining region, as well as the location of historic gold mines. 8.5" X 11", 156 ppgs. Retail Price: $19.99

Mines of Plumas County, California - Unavailable since 1918, this publication was originally compiled by the California State Mining Bureau and includes details on the gold mines of Plumas County, California. Also included are insights into the mineralization and other characteristics of this important mining region, as well as the location of historic gold mines. 8.5" X 11", 200 ppgs. Retail Price: $19.99

Mines of El Dorado, Placer, Sacramento and Yuba Counties, California - Originally published in 1917, this important publication on California Mining has not been available for nearly a century. This publication includes important information on the early gold mines of El Dorado County, Placer County, Sacramento County and Yuba County, which were some of the first gold fields mined by the Forty-Niners during the California Gold Rush. Included are not only details on early mining methods in the area, production statistics and geological information, but also the location of the early gold mines that helped make California "The Golden State". Also included are insights into the early mining of chrome, copper and other minerals in this important mining area. 8.5" X 11", 204 ppgs. Retail Price: $19.99

Mines of Los Angeles, Orange and Riverside Counties, California - Originally published in 1917, this important publication on California Mining has not been available for nearly a century. This publication includes important information on the early gold mines of Los Angeles County, Orange County and Riverside County, which were some of the first gold fields mined in California by early Spanish and Mexican miners before the 49ers came on the scene. Included are not only details on early mining methods in the area, production statistics and geological information, but also the location of the early gold mines that helped make California "The Golden State". 8.5" X 11", 146 ppgs. Retail Price: $12.99

Mines of San Bernadino and Tulare Counties, California - Originally published in 1917, this important publication on California Mining has not been available for nearly a century. This publication includes important information on the early gold mines of San Bernadino and Tulare County, which were some of the first gold fields mined in California by early Spanish and Mexican miners before the 49ers came on the scene. Included are not only details on early mining methods in the area, production statistics and geological information, but also the location of the early gold mines that helped make California "The Golden State". Also included are details on the mining of other minerals such as copper, iron, lead, zinc, manganese, tungsten, vanadium, asbestos, barite, borax, cement, clay, dolomite, fluospar, gem stones, graphite, marble, salines, petroleum, stronium, talc and others. 8.5" X 11", 200 ppgs. Retail Price: $19.99

Chromite Mining in The Klamath Mountains of California and Oregon - Unavailable since 1919, this publication was originally compiled by J.S. Diller of the United States Department of Geological Survey and includes details on the chromite mines of this region of Northern California and Southern Oregon. Also included are insights into the mineralization and other characteristics of this important mining region, as well as the location of historic mines. Also included are insights into chromite mining in Eastern Oregon and Montana. 8.5" X 11", 98 ppgs. Retail Price: $9.99

Mines and Mining in Amador, Calaveras and Tuolumne Counties, California - Unavailable since 1915, this publication was originally compiled by William Tucker and includes details on the mines and mineral resources of this important California mining area. Included are details on the geology, history and important gold mines of the region, as well as insights into other local mineral resources such as asbestos, clay, copper, talc, limestone and others. Also included are insights into the mineralization and other characteristics of this important portion of California's Mother Lode mining region. 8.5" X 11", 198 ppgs. Retail Price: $14.99

The Cerro Gordo Mining District of Inyo County California - Unavailable since 1963, this publication was originally compiled by the United States Department of Interior. Included are insights into the mineralization and other characteristics of this important mining region of Southern California. Topics include the mining of gold and silver in this important mining district in Inyo County, California, including details on the history, production and locations of the Cerro Gordo Mine, the Morning Star Mine, Estelle Tunnel, Charles Lease Tunnel, Ignacio, Hart, Crosscut Tunnel, Sunset, Upper Newtown, Newtown, Ella, Perseverance, Newsboy, Belmont and other silver and gold mines in the Cerro Gordo Mining District. This volume also includes important insights into the fossil record, geologic formations, faults and other aspects of economic geology in this California mining district. 8.5" X 11", 104 ppgs. Retail Price: $10.99

Mining in Butte, Lassen, Modoc, Sutter and Tehama Counties of California - Unavailable since 1917, this publication was originally compiled by the United States Department of Interior. Included are insights into the mineralization and other characteristics of this important mining region of California. Topics include the mining of asbestos, chromite, gold, diamonds and manganese in Butte County, the mining of gold and copper in the Hayden Hill and Diamond Mountain mining districts of Lassen County, the mining of coal, salt, copper and gold in the High Grade and Winters mining districts of Modoc County, gold mining in Sutter County and the mining of gold, chromite, manganese and copper in Tehama County. This volume also includes the production records and locations of numerous mines in this important mining region. 8.5" X 11", 114 ppgs. Retail Price: $11.99

Mines of Trinity County California - Originally published in 1965, this important publication on California Mining has not been available for nearly fifty years. This publication includes important information on mines and mining in Trinity County, California, as well insights into the mineralization and geology of this important mining area in Northern California. Included are extensive details on hardrock and placer gold mines and prospects, including charts showing the locations of these historic mines.. 8.5" X 11", 144 ppgs. Retail Price: $12.99

Mines of Kern County California - Originally published in 1962, this important publication on California Mining has not been available for nearly fifty years. This publication includes important information on mines and mining in Kern County, California, as well insights into the mineralization and geology of this important mining area in California. Included are extensive details on hardrock and placer gold mines and prospects, including charts showing the locations of these historic mines. 8.5" X 11", 398 ppgs. Retail Price: $24.99

Mines of Calaveras County California - Originally published in 1962, this important publication on California Mining has not been available for nearly fifty years. This publication includes important information on mines and mining in Calaveras County, California, as well insights into the mineralization and geology of this important mining area in Northern California. Included are extensive details on hardrock and placer gold mines and prospects, including charts showing the locations of these historic mines. 8.5" X 11", 236 ppgs. Retail Price: $19.99

Lode Gold Mining in Grass Valley California - Unavailable since 1940, this publication was originally compiled by the United States Department of Interior. Included are insights into the gold mineralization and other characteristics of this important mining region of Nevada County, California. This volume also includes important insights into the geologic formations, faults and other aspects of economic geology in this California mining district. Of particular interest are the fine details on many hardrock gold mines in the area, including their locations, histories, development and mineralization. Some of the mines featured include the Gold Hill Mine, Massachusetts Hill, Boundary, Peabody, Golden Center, North Star, Omaha, Lone Jack, Homeward Bound, Hartery, Wisconsin, Allison Ranch, Phoenix, Kate Hayes, W.Y.O.D., Empire, Rich Hill, Daisy Hill, Orleans, Sultana, Centennial, Conlin, Ben Franklin, Crown Point and many others. 8.5" X 11", 148 ppgs. Retail Price: $12.99

Lode Mining in the Alleghany District of Sierra County California - Unavailable since 1913, this publication was originally compiled by the United States Department of Interior. Included are insights into the mineralization and other characteristics of this important mining region of Sierra County. Included are details on the history, production and locations of numerous hardrock gold mines in this famous California area, including the Tightner Mine, Minnie D., Osceola, Eldorado, Twenty One, Sherman, Kenton, Oriental, Rainbow, Plumbago, Irelan, Gold Canyon, North Fork, Federal, Kate Hardy and others. This volume also includes important insights into the fossil record, geologic formations, faults and other aspects of economic geology in this California mining district. 8.5" X 11", 48 ppgs. Retail Price: $7.99

Six Months In The Gold Mines During The California Gold Rush - Unavailable since 1850, this important work is a first hand account of one "49'ers" personal experience during the great California Gold Rush, shedding important light on one of the most exciting periods in the history of not only California, but also the world. Compiled from journals written between 1847 and 1849 by E. Gould Buffum, a native of New York, "Six Months In The Gold Mines During The California Gold Rush" offers a rare look into the day to day lives of the people who came to California to work in her gold mines when the state was still a great frontier. 8.5" X 11", 290 ppgs. Retail Price: $19.99

Quartz Mines of the Grass Valley Mining District of California - Unavailable since 1867, this important publication has not been available since those days. This rare publication offers a short dissertation on the early hardrock mines in this important mining district in the California Mother Lode region between the 1850's and 1860's. Also included are hard to find details on the mineralization and locations of these mines, as well as how they were operated in those day. 8.5" X 11", 44 ppgs. Retail Price: $8.99

Gold Rush on the Feather River - First published in 1924, this short publication by G.C. Mansfield sheds important light on the early history of gold mining on the Feather River. Included are rare insights into the first decade of gold mining and the early mining camps of the Feather River during the 1850's. 64 ppgs., 9.99

The Bodie Mining District of California - First published in 1986, it has been unavailable since those days and sheds important light on this famous mining area. Included are the history, characteristics and locations of numerous old mines around the ghost town of Bodie. 64 ppgs, 8.99

Geology and Mineral Resources of the Gasquet Quadrangle of California-Oregon - First published in 1953, it has been unavailable for over a century and sheds important light on the geological features and mineral resources of this portion of Northern California and Southern Oregon. 80 ppgs, 9.99

Alaska Mining Books

Ore Deposits of the Willow Creek Mining District, Alaska - Unavailable since 1954, this hard to find publication includes valuable insights into the Willow Creek Mining District near Hatcher Pass in Alaska. The publication includes insights into the history, geology and locations of the well known mines in the area, including the Gold Cord, Independence, Fern, Mabel, Lonesome, Snowbird, Schroff-O'Neil, High Grade, Marion Twin, Thorpe, Webfoot, Kelly-Willow, Lane, Holland and others. 8.5" X 11", 96 ppgs. Retail Price: $9.99

The Juneau Gold Belt of Alaska - Unavailable since 1906, this hard to find publication includes valuable insights into the gold mines around Juneau, Alaska. The publication includes important details into the history, geology and locations of the well known gold mines and prospects in the area, including those around Windham Bay, Holkham Bay, Port Snettisham, on Grindstone and Rhine Creeks, Gold Creek, Douglas Island, Salmon Creek, Lemon Creek, Nugget Creek, from the Mendenhall River to Berners Bay, McGinnis Creek, Montana Creek, Peterson Creek, Windfall Creek, the Eagle River, Yankee Basin, Yankee Curve, Kowee Creek and elsewhere. Not only are gold placer mines included, but also hardrock gold mines. 8.5" X 11", 224 ppgs. Retail Price: $19.99

Mining in the Jumbo Basin of Alaska - Unavailable since 1953, this hard to find publication includes valuable insights into the mines and geology of the Jumbo Basin. The publication includes important details into the history, geology and locations of the well known gold mines and prospects in the famous Jumbo Basin Mining Region of Alaska. 72 ppgs, 9.99

The Rampart Placer Gold Region of Alaska - Unavailable since 1906, this hard to find publication includes valuable insights into the placer gold mines of the Rampart Mining Region. The publication includes important details into the history, geology and locations of the well known gold mines and prospects in the famous Rampart Mining Region of Alaska. 78 ppgs, 10.99

Arizona Mining Books

Mines and Mining in Northern Yuma County Arizona - Originally published in 1911, this important publication on Arizona Mining has not been available for over a hundred years. Included are rare insights into the gold, silver, copper and quicksilver mines of Yuma County, Arizona together with hard to find maps and photographs. Some of the mines and mining districts featured include the Planet Copper Mine, Mineral Hill, the Clara Consolidated Mine, Viati Mine, Copper Basin prospect, Bowman Mine, Quartz King, Billy Mack, Carnation, the Wardwell and Osbourne, Valensuella Copper, the Mariquita, Colonial Mine, the French American, the New York-Plomosa, Guadalupe, Lead Camp, Mudersbach Copper Camp, Yellow Bird, the Arizona Northern (Salome Strike), Bonanza (Harqua Hala), Golden Eagle, Hercules, Socorro and others. 8.5" X 11", 144 ppgs. Retail Price: $11.99

The Aravaipa and Stanley Mining Districts of Graham County Arizona - Originally published in 1925, this important publication on Arizona Mining has not been available for nearly ninety years. Included are rare insights into the gold and silver mines of these two important mining districts, together with hard to find maps. 8.5" X 11", 140 ppgs. Retail Price: $11.99

Gold in the Gold Basin and Lost Basin Mining Districts of Mohave County, Arizona - This volume contains rare insights into the geology and gold mineralization of the Gold Basin and Lost Basin Mining Districts of Mohave County, Arizona that will be of benefit to miners and prospectors. Also included is a significant body of information on the gold mines and prospects of this portion of Arizona. This volume is lavishly illustrated with rare photos and mining maps. **8.5" X 11", 188 ppgs. Retail Price: $19.99**

Mines of the Jerome and Bradshaw Mountains of Arizona - This important publication on Arizona Mining has not been available for ninety years. This volume contains rare insights into the geology and ore deposits of the Jerome and Bradshaw Mountains of Arizona that will be of benefit to miners and prospectors who work those areas. Included is a significant body of information on the mines and prospects of the Verde, Black Hills, Cherry Creek, Prescott, Walker, Groom Creek, Hassayampa, Bigbug, Turkey Creek, Agua Fria, Black Canyon, Peck, Tiger, Pine Grove, Bradshaw, Tintop, Humbug and Castle Creek Mining Districts. This volume is lavishly illustrated with rare photos and mining maps. **8.5" X 11", 218 ppgs. Retail Price: $19.99**

The Ajo Mining District of Pima County Arizona - This important publication on Arizona Mining has not been available for nearly seventy years. This volume contains rare insights into the geology and mineralization of the Ajo Mining District in Pima County, Arizona and in particular the famous New Cornelia Mine. **8.5" X 11", 126 ppgs. Retail Price: $11.99**

Mining in the Santa Rita and Patagonia Mountains of Arizona - Originally published in 1915, this important publication on Arizona Mining has not been available for nearly a century. Included are rare insights into hundreds of gold, silver, copper and other mines in this famous Arizona mining area. Details include the locations, geology, history, production and other facts of the mines of this region. **8.5" X 11", 394 ppgs. Retail Price: $24.99**

Mining in the Bisbee Quadrangle of Arizona - Originally published in 1906, this important publication on Arizona Mining has not been available for nearly a century. Included are rare insights into hundreds of gold, silver, copper and other mines in this famous Arizona mining area. Details include the locations, geology, history, production and other facts of the mines of this important mining region. **8.5" X 11", 188 ppgs. Retail Price: $14.99**

Placer Gold Mining in Arizona - Unavailable since 1922, this hard to find publication includes valuable insights into the placer gold mines of the Arizona. Originally released as "Placer Gold of Arizona", despite its small size, this publication includes important details into the history, geology and locations of the well known placer gold mines and prospects in the State of Arizona. **48 ppgs, 8.99**

Gold and Copper Mining near Payson, Arizona - Written in 1915, this hard to find publication includes valuable insights into the gold and copper mining industry of Arizona. Highlighted here are the gold and copper mines near Payson, Arizona. **68 ppgs, 8.99**

Lode Gold Mining in Arizona - Unavailable since 1934, this hard to find publication, originally released as "Arizona Lode Gold Mines and Gold Mining" includes valuable insights into the gold mining industry of Arizona. Included are valuable insights into over 150 hardrock gold mines in over 30 different mining districts in Arizona. **278 ppgs, 21.99**

Mining in the Dragoon Quadrangle of Cochise County, Arizona - Unavailable since 1964, this hard to find publication includes valuable insights into the mines of the Dragoon Quadrangle Mining Region. The publication includes important details into the history, geology and locations of the well known mines and prospects in this famous mining region of Arizona. **224 ppgs., 19.99**

Directory of Operating Mines in Arizona in 1915 - Unavailable since 1916, this hard to find publication includes valuable insights into the mines of Arizona. This small publication includes a complete list of the mines that were operating in the State of Arizona during 1915 and includes details such as general location, owners and some basic facts about each mining operation. **52 ppgs. 8.99**

Arizona Ore Deposits - Unavailable since 1938, this hard to find publication includes valuable insights into some ore deposits of Arizona. Included are valuable insights into the formation and characteristics of valuable ore deposits in the Jerome, Miami, Inspiration, Clifton, Morenci, Ray, Ajo, Eureka, Tombstone and Magma mining districts. Included are details into some of the major gold, silver and copper mines of these important Arizona mining areas. **160 ppgs, 14.99**

Montana Mining Books

A History of Butte Montana: The World's Greatest Mining Camp - First published in 1900 by H.C. Freeman, this important publication sheds a bright light on one of the most important mining areas in the history of The West. Together with his insights, as well as rare photographs of the periods, Harry Freeman describes Butte and its vicinity from its early beginnings, right up to its flush years when copper flowed from its mines like a river. At the time of publication, Butte, Montana was known worldwide as "The Richest Mining Spot On Earth" and produced not only vast amounts of copper, but also silver, gold and other metals from its mines. Freeman illustrates, with great detail, the most important mines in the vicinity of Butte, providing rare details on their owners, their history and most importantly, how the mines operated and how their treasures were extracted. Of particular interest are the dozens of rare photographs that depict mines such as the famous Anaconda, the Silver Bow, the Smoke House, Moose, Paulin, Buffalo, Little Minah, the Mountain Consolidated, West Greyrock, Cora, the Green Mountain, Diamond, Bell, Parnell, the Neversweat, Nipper, Original and many others. **8.5" X 11", 142 ppgs. Retail Price: $12.99**

The Butte Mining District of Montana - This important publication on Montana Mining has not been available for over a century. Included are rare insights into the gold, copper and silver mines of Butte, Montana together with hard to find maps and photographs. Some of the topics include the early history of gold, silver and copper mining in the Butte area, insight into the geology of its mining areas, the local distribution of gold, silver and copper ores, as well their composition and how to identify them. Also included are detailed facts about the mines in the Butte Mining District, including the famous Anaconda Mine, Gagnon, Parrot, Blue Vein, Moscow, Poulin, Stella, Buffalo, Green Mountain, Wake Up Jim, the Diamond-Bell Group, Mountain Consolidated, East Greyrock, West Greyrock, Snowball, Corra, Speculator, Adirondack, Miners Union, the Jessie-Edith May Group, Otisco, Iduna, Colorado, Lizzie, Cambers, Anderson, Hesperus, Preferencia and dozens of others. **8.5" X 11", 298 ppgs. Retail Price: $24.99**

Mines of the Helena Mining Region of Montana - This important publication on Montana Mining has not been available for over a century. Included are rare insights into the gold, copper and silver mines of the vicinity of Helena, Montana, including the Marysville Mining District, Elliston Mining District, Rimini Mining District, Helena Mining District, Clancy Mining District, Wickes Mining District, Boulder and Basin Mining Districts and the Elkhorn Mining District. Some of the topics include the early history of gold, silver and copper mining in the Helena area, insight into the geology of its mining areas, the local distribution of gold, silver and copper ores, as well their composition and how to identify them. Also included are detailed facts, history, geology and locations of over one hundred gold, silver and copper mines in the area . **8.5" X 11", 162 ppgs, Retail Price: $14.99**

Mines and Geology of the Garnet Range of Montana - This important publication on Montana Mining has not been available for over a century. Included are rare insights into the gold, copper and silver mines of the vicinity of this important mining area of Montana. Some of the topics include the early history of gold, silver and copper mining in the Garnet Mountains, insight into the geology of its mining areas, the local distribution of gold, silver and copper ores, as well their composition and how to identify them. Also included are detailed facts, history, geology and locations of numerous gold, silver and copper mines in the area . **8.5" X 11", 100 ppgs, Retail Price: $11.99**

Mines and Geology of the Philipsburg Quadrangle of Montana - This important publication on Montana Mining has not been available for over a century. Included are rare insights into the gold, copper and silver mines of the vicinity of this important mining area of Montana. Some of the topics include the early history of gold, silver and copper mining in the Philipsburg Quadrangle, insight into the geology of its mining areas, the local distribution of gold, silver and copper ores, as well their composition and how to identify them. Also included are detailed facts, history, geology and locations of over one hundred gold, silver and copper mines in the area **8.5" X 11", 290 ppgs, Retail Price: $24.99**

Geology of the Marysville Mining District of Montana - Included are rare insights into the mining geology of the Marysville Mining District. Some of the topics include the early history of gold, silver and copper mining in the area, insight into the geology of its mining areas, the local distribution of gold, silver and copper ores, as well their composition and how to identify them. Also included are detailed facts, history, geology and locations of gold, silver and copper mines in the area **8.5" X 11", 198 ppgs, Retail Price: $19.99**

The Geology and Mines of Northern Idaho and North Western Montana- See listing under Idaho.

The History of Gold Dredging in Montana - Unavailable since 1916, this important publication was originally published by the Us Bureau of Mines and has been unavailable for a century. A century and more ago, giant dredging machines dug in Montana's rivers and creeks in search of illusive golden riches. First appearing in California in the 1850's, gold dredges finally reached their peak of development in Siberia and New Zealand before becoming popular again in the United States. This book offers a unique historical perspective on the gold dredges that once operated in Montana. This book on Montana mining history is lavishly illustrated with dozens of rare historic photos gold dredges that once operated in Montana, as well as hard to locate plans on how these dredges were designed. 120 ppgs., 11.99

Nevada Mining Books

The Bull Frog Mining District of Nevada - Unavailable since 1910, this publication was originally compiled by the United States Department of Interior. This volume also includes important insights into the geologic formations, faults and other aspects of economic geology in this Nevada mining district. Of particular interest are the fine details on many mines in the area, including their locations, histories, development and mineralization. Some of the mines featured include the National Bank Mine, Providence, Gibraltor, Tramps, Denver, Original Bullfrog, Gold Bar, Mayflower, Homestake-King and other mines and prospects. 8.5" X 11", 152 ppgs, Retail Price: $14.99

History of the Comstock Lode - Unavailable since 1876, this publication was originally released by John Wiley & Sons. This volume also includes important insights into the famous Comstock Lode of Nevada that represented the first major silver discovery in the United States. During its spectacular run, the Comstock produced over 192 million ounces of silver and 8.2 million ounces of gold. Not only did the Comstock result in one of the largest mining rushes in history and yield immense fortunes for its owners, but it made important contributions to the development of the State of Nevada, as well as neighboring California. Included here are important details on not only the early development and history of the Comstock, but also rare early insight into its mines, ore and its geology. 8.5" X 11", 244 ppgs, Retail Price: $19.99

The Pioche Mining District of Nevada - First published in 1932, it has been unavailable for over a century and sheds important light on the mining history of Nevada. Some of the topics include the history of mining in this district, as well as the characteristics of its mineral and ore deposits. Also included are insights into the history, production, characteristics and locations of numerous mines in the area. Some of the mines include the Combined Metals, Pioche, Ely Valley, No. 10, Poorman, Wide Awake, Alps, Prince, Virginia Louise, Half Moon, Abe Lincoln, Fairview, Bristol Silver, National, Vesuvius, Inman, Tempest, Hillside, Jackrabbit, Lucky Star, Fortuna, Mendha, Manhattan, Hamburg, Comet, Lyndon and others. 108 ppgs 10.99

The Yerington Mining District of Nevada - First published in 1932, it has been unavailable for over a century and sheds important light on the mining history of Nevada. Some of the topics include the history of mining in this district, as well as the characteristics of its mineral and ore deposits. Also included are insights into the history, production, characteristics and locations of numerous mines in the area. Some of the mines include the Bluestone, Mason Valley, Malachite, McConnell, Greenwood, Western Nevada, Ludwig, Douglas Hill, Casting Copper, Montana-Yerington, Empire, Jim Beatty, Terry and McFarland, Blue Jay and others. 92 ppgs, 10.99

The Genesis of the Ores of Tonopah Nevada - Unavailable since 1918, this hard to find publication includes valuable insights into the gold mines around Tonopah, Nevada. The publication includes important details into the geology of mines in the Tonopah Mining District of Nevada. 90 ppgs, 10.99

Mining Camps of Elko, Lander and Eureka Counties Nevada - Unavailable since 1910, this hard to find publication includes valuable insights into the mining camps of Elko, Lander and Eureka Counties, Nevada. The publication includes important details into the history of mines and mining in these three Nevada counties. 154 ppgs, 12.99

Ore Deposits of the Bullfrog Quadrangle - Unavailable since 1964 and released as "Geology of Bullfrog Quadrangle and Ore Deposits Related to Bullfrog Hills Caldera, Nye County, Nevada and Inyo County, California". The publication includes important details into the geology of mines in the Bullfrog Quadrangle of Nye County, Nevada and Inyo County, California. 52 ppgs, 9.99

Mining in Eureka County Nevada - Unavailable since 1879, this hard to find publication includes valuable insights into the early mining history off Eureka County, Nevada. The publication includes important details into the early history of the mines of Eureka County, as well as their development, production and how their ores were treated. Also included are details on the 1872 Mining Act, as well as the local rules, regulations and customs of the miners in Eureka County. 134 ppgs, 12.99

Colorado Mining Books

Ores of The Leadville Mining District - Unavailable since 1926, this publication was originally compiled by the United States Department of Interior. This volume also includes important insights into the ores and mineralization of the Leadville Mining District in Colorado. Topics include historic ore prospecting methods, local geology, insights into ore veins and stockworks, the local trend and distribution of ore channels, reverse faults, shattered rock above replacement ore bodies, mineral enrichment in oxidized and sulphide zones and more. **8.5″ X 11″, 66 ppgs, Retail Price: $8.99**

Mining in Colorado - Unavailable since 1926, this publication was originally compiled by the United States Department of Interior. This volume also includes important insights into the mining history of Colorado from its early beginnings in the 1850's right up to the mid 1920's. Not only is Colorado's gold mining heritage included, but also its silver, copper, lead and zinc mining industry. Each mining area is treated separately, detailing the development of Colorado's mines on a county by county basis. **8.5″ X 11″, 284 ppgs, Retail Price: $19.99**

Gold Mining in Gilpin County Colorado - Unavailable since 1876, this publication was originally compiled by the Register Steam Printing House of Central City, Colorado. A rare glimpse at the gold mining history and early mines of Gilpin County, Colorado from their first discovery in the 1850's up to the "flush years" of the mid 1870's. Of particular interest is the history of the discovery of gold in Gilpin County and details about the men who made those first strikes. Special focus is given to the early gold mines and first mining districts of the area, many of which are not detailed in other books on Colorado's gold mining history. **8.5″ X 11″, 156 ppgs, Retail Price: $12.99**

Mining in the Gold Brick Mining District of Colorado - Important insights into the history of the Gold Brick Mining District, as well as its local geography and economic geology. Also included are the histories and locations of historic mines in this important Colorado Mining District, including the Cortland, Carter, Raymond, Gold Links, Sacramento, Bassick, Sandy Hook, Chronicle, Grand Prize, Chloride, Granite Mountain, Lucille, Gray Mountain, Hilltop, Maggie Mitchell, Silver Islet, Revenue, Roosevelt, Carbonate King and others. In addition to hardrock mining, are also included are details on gold placer mining in this portion of Colorado. **8.5″ X 11″, 140 ppgs, Retail Price: $12.99**

Ore Deposits of the London Fault of Colorado - First published in 1941, it has been unavailable since those days and sheds important light on the mines and mineral deposits of the London Fault in Central Colorado's Alma Mining District. This publication sheds important light on the gold veins and lead-silver deposits of the Alma Mining District. Included are geologic details on the London Mine, American Mine, Havigorst Tunnel, Ophir Mine, Mosher Tunnel, London-Butte Mine, Venture Shaft, Hard-To-Beat Mine, Oliver Twist Tunnel, Sacramento Mine, Mudsill Mine, Sherwood Mine, Wagner, Barcoe Tunnel and other mines in this important mining region. 110 ppgs., 10.99

The Mines of Colorado - First published in 1867, it has been unavailable since those days and sheds important light on Colorado's early mining history. Written shortly after the events took place, this publication sheds important light on the Pike's Peak Gold Rush, the discovery of gold on Ralston Creek and Dry Creek in the 1850's, as well as details on the first wave of miners into Colorado and their trials and tribulations as they crossed the Great Plains. Also included are details on early discoveries of lode gold in the mountainous regions of Colorado, details on the early mines hardrock and placer mines, and much more. It is a veritable treasure trove on Colorado's early mining history and will be of great importance to anyone who is interested in the mining of gold or other minerals in Colorado, as well as those interested in the history of the state. 478 ppgs., 29.99

The La Plata Mining District of Colorado - Originally titled "Geology and Ore Deposits in the Vicinity of the La Plata District of Colorado" and first published in 1949, it has been unavailable since those days and sheds important light on the mines and mineral deposits of the La Plata Mining District of Colorado. 214 ppgs., 19.99

Washington Mining Books

The Republic Mining District of Washington - Unavailable since 1910, this important publication was originally published by the Washington Geologic Survey and has been unavailable for a century. Topics include the geology, rock formations and the formation of ore deposits in this important mining area of Washington State. Also included are hard to find details on the geology, history and locations of dozens of mines in the area. Some of the mines featured include the New Republic Mine, Ben Hur, Morning Glory, the South Republic Mine, Quilp, Surprise, Black Tail, Lone Pine, San Poil, Mountain Lion, Tom Thumb, Elcaliph and many others. **8.5" X 11", 94 ppgs, Retail Price: $10.99**

The Myers Creek and Nighthawk Mining Districts of Washington - Unavailable since 1911, this important publication was originally published by the Washington Geologic Survey and has been unavailable for a century. Topics include the geology, rock formations and the formation of ore deposits in these important mining areas of Washington State. Also included are hard to find details on the geology, history and locations of dozens of mines in the area. Some of the mines featured include the Grant Mine, Monterey, Nip and Tuck, Myers Creek, Number Nine, Neutral, Rainbow, Aztec, Crystal Butte, Apex, Butcher Boy, Molson, Mad River, Olentangy, Delate, Kelsey, Golden Chariot, Okanogan, Ohio, Forty-Ninth Parallel, Nighthawk, Favorite, Little Chopaka, Summit, Number One, California, Peerless, Caaba, Prize Group, Ruby, Mountain Sheep, Golden Zone, Rich Bar, Similkameen, Kimberly, Triune, Hiawatha, Trinity, Hornsilver, Maquae, Bellevue, Bullfrog, Palmer Lake, Ivanhoe, Copper World and many others. **8.5" X 11", 136 ppgs, Retail Price: $12.99**

The Blewett Mining District of Washington - Unavailable since 1911, this important publication was originally published by the Washington Geologic Survey and has been unavailable for a century. Topics include the geology, rock formations and the formation of ore deposits in this important mining area of Washington State. Also included are hard to find details on the geology, history and locations of dozens of mines in the area. Some of the mines featured include the Washington Meteor, Alta Vista, Pole Pick, Blinn, North Star, Golden Eagle, Tip Top, Wilder, Golden Guinea, Lucky Queen, Blue Bell, Prospect, Homestake, Lone Rock, Johnson, and others. **8.5" X 11", 134 ppgs, Retail Price: $12.99**

Silver Mining In Washington - Unavailable since 1955, this important publication was originally published by the Washington Geologic Survey. Featured are the hard to find locations and details pertaining to Washington's silver mines. **8.5" X 11", 180 ppgs, Retail Price: $15.99**

The Mines of Snohomish County Washington - Unavailable since 1942, this important publication was originally published by the Washington Geologic Survey and has been unavailable for seventy years. Featured are details on a large number of gold, silver, copper, lead and other metallic mineral mines. Included are the locations of each historic mine, along with information on the commodity produced. **8.5" X 11", 98 ppgs, Retail Price: $10.99**

The Mines of Chelan County Washington - Unavailable since 1943, this important publication was originally published by the Washington Geologic Survey and has been unavailable for seventy years. Featured are details on a large number of gold, silver, copper, lead and other metallic mineral mines. Included are the locations of each historic mine, along with information on the commodity. **8.5" X 11", 88 ppgs, Retail Price: $9.99**

Metal Mines of Washington - Unavailable since 1921, this important publication was originally published by the Washington Geologic Survey and has been unavailable for nearly ninety years. Widely considered a masterpiece on the Washington Mining Industry, "Metal Mines of Washington" sheds light on the important details of Washington's early mining years. Featured are details on hundreds of gold, silver, copper, lead and other metallic mineral mines. Included are hard to find details on the mineral resources of this state, as well as the locations of historic mines. Lavishly illustrated with maps and historic photos and complete with a glossary to explain any technical terms found in the text, this is one of the most important works on mining in the State of Washington. No prospector or miner should be without it if they are interested in mining in Washington. **8.5" X 11", 396 ppgs, Retail Price: $24.99**

Gem Stones In Washington - Unavailable since 1949, this important publication was originally published by the Washington Geologic Survey and has been unavailable since first published. Included are details on where to find naturally occurring gem stones in the State of Washington, including quartz crystal, amethyst, smoky quartz, milky quartz, agates, bloodstone, carnelian, chert, flint, jasper, onyx, petrified wood, opal, fire opal, hyalite and others. **8.5" X 11", 54 ppgs, Retail Price: $8.99**

The Covada Mining District of Washington - Unavailable since 1913, this important publication was originally published by the Washington Geologic Survey and has been unavailable for a century. Topics include the geology, rock formations and the formation of ore deposits in this important mining area of Washington State. Also included are hard to find details on the geology, history and locations of dozens of mines in the area. Some of the mines featured include the Admiral, Advance, Algonkian, Big Bug, Big Chief, Big Joker, Black Hawk, Black Tail, Black Thorn, Captain, Cherokee Strip, Colorado, Dan Patch, Dead Shot, Etta, Good Ore, Greasy Run, Great Scott, Idora, IXL, Jay Bird, Kentucky Bell, King Solomon, Laurel, Laura S, Little Jay, Meteor, Neglected, Northern Light, Old Nell, Plymouth Rock, Polaris, Quandary, Reserve, Shoo Fly, Silver Plume, Three Pines, Vernie, White Rose and dozens of others. **8.5" X 11", 114 ppgs, Retail Price: $10.99**

The Index Mining District of Washington - Unavailable since 1912, this important publication was originally published by the Washington Geologic Survey and has been unavailable for a century. Topics include the geology, rock formations and the formation of ore deposits in this important mining area of Washington State. Also included are hard to find details on the geology, history and locations of dozens of mines in the area. Some of the mines featured include the Sunset, Non-Pareil, Ethel Consolidated, Kittaning, Merchant, Homestead, Co-operative, Lost Creek, Uncle Sam, Calumet, Florence-Rae, Bitter Creek, Index Peacock, Gunn Peak, Helena, North Star, Buckeye. Copper Bell, Red Cross and others. **8.5" X 11", 114 ppgs, Retail Price: $11.99**

Mining & Mineral Resources of Stevens County Washington - Unavailable since 1920, this important publication was originally published by the Washington Geologic Survey and has been unavailable for a century. Topics include the geology, rock formations and the formation of ore deposits in these important mining areas of Washington State. Also included are hard to find details on the geology, history and locations of hundreds of mines in the area. **8.5" X 11", 372 ppgs, Retail Price: $24.99**

The Mines and Geology of the Loomis Quadrangle Okanogan County, Washington - Unavailable since 1972, this important publication was originally published by the Washington Geologic Survey and has been unavailable for a century. Topics include the geology, rock formations and the formation of ore deposits in this important mining area of Washington State. Also included are hard to find details on the geology, history and locations of dozens of gold, copper, silver and other mines in the area. **8.5" X 11", 150 ppgs, Retail Price: $12.99**

The Conconully Mining District of Okanogan County Washington - Unavailable since 1973, this important publication was originally published by the Washington Geologic Survey and has been unavailable for a century. Topics include the geology, rock formations and the formation of ore deposits in this important mining area of Washington State, which also includes Salmon Creek, Blue Lake and Galena. Also included are hard to find details on the geology, mining history and locations of dozens of mines in the area. Some of the mines include Arlington, Fourth of July, Sonny Boy, First Thought, Last Chance, War Eagle-Peacock, Wheeler, Mohawk, Lone Star, Woo Loo Moo Loo, Keystone, Hughes, Plant-Callahan, Johnny Boy, Leuena, Gubser, John Arthur, Tough Nut, Homestake, Key and many others **8.5" X 11", 68 ppgs, Retail Price: $8.99**

Wyoming Mining Books

Mining in the Laramie Basin of Wyoming - Unavailable since 1909, this publication was originally compiled by the United States Department of Interior. Also included are insights into the mineralization and other characteristics of this important mining region, especially in regards to coal, limestone, gypsum, bentonite clay, cement, sand, clay and copper. **8.5" X 11", 104 ppgs, Retail Price: $11.99**

New Mexico Mining Books

The Mogollon Mining District of New Mexico - Unavailable since 1927, this important publication was originally published by the US Department of Interior and has been unavailable for 80 years. Topics include the geology, rock formations and the formation of ore deposits in this important mining area in New Mexico. Of particular focus is information on the history and production of the ore deposits in this area, their form and structure, vein filling, their paragenesis, origins and ore shoots, as well as oxidation and supergene enrichment. Also included are hard to find details, including the descriptions and locations of numerous gold, silver and other types of mines, including the Eureka, Pacific, South Alpine, Great Western, Enterprise, Buffalo, Mountain View, Floride, Gold Dust, Last Chance, Deadwood, Confidence, Maud S., Deep Down, Little Fanney, Trilby, Johnson, Alberta, Comet, Golden Eagle, Cooney, Queen, the Iron Crown, Eberle, Clifton, Andrew Jackson mine, Mascot and others. **8.5" X 11", 144 ppgs, Retail Price: $12.99**

The Percha Mining District of Kingston New Mexico - Unavailable since 1883, this important publication was originally published by the Kingston Tribune and has been unavailable for over one hundred and thirty five years. Having been written during the earliest years of gold and silver mining in the Percha Mining District, unlike other books on the subject, this work offers the unique perspective of having actually been written while the early mining history of this area was still being made. In fact, the work was written so early in the development of this area that many of the notable mines in the Percha District were less than a few years old and were still being operated by their original discoverers with the same enthusiasm as when they were first located. Included are hard to find details on the very earliest gold and silver mines of this important mining district near Kingston in Sierra County, New Mexico. **8.5" X 11", 68 ppgs, Retail Price: $9.99**

East Coast Mining Books

<u>The Gold Fields of the Southern Appalachians</u> - Unavailable since 1895, this important publication was originally published by the US Department of Interior and has been unavailable for nearly 120 years. Topics include the geology, rock formations and the formation of ore deposits in this important mining area of the American South. Of particular focus is information on the history and statistics of the ore deposits in this area, their form and structure and veins. Also included are details on the placer gold deposits of the region. The gold fields of the Georgian Belt, Carolinian Belt and the South Mountain Mining District of North Carolina are all treated in descriptive detail. Included are hard to find details, including the descriptions and locations of numerous gold mines in Georgia, North Carolina and elsewhere in the American South. Also included are details on the gold belts of the British Maritime Provinces and the Green Mountains. **8.5" X 11", 104 ppgs, Retail Price: $9.99**

Gold Rush Tales Series

<u>Millions in Siskiyou County Gold</u> - In this first volume of the "Gold Rush Tales" series, leading mining historian and editor Kerby Jackson, introduces us to the story of how millions of dollars worth of gold was discovered in Siskiyou County during the California Gold Rush. Lavishly illustrated with photos from the 19th Century, this hard to find information was first published in 1897 and sheds important light onto the gold rush era in Siskiyou County, California and the experiences of the men who dug for the gold and actually found it. **8.5" X 11", 82 ppgs, Retail Price: $9.99**

<u>The California Rand in the Days of '49</u> - In this second volume of the "Gold Rush Tales" series, leading mining historian and editor Kerby Jackson, introduces us to four tales from the California Gold Rush. Lavishly illustrated with photos from the 19th Century, this hard to find information was first published in 1890's and includes the stories of "California's Rand", details about Chinese miners, how one early miner named Baker struck it rich and also the story of Alphonzo Bowers, who invented the first hydraulic gold dredge. **8.5" X 11", 54 ppgs, Retail Price: $9.99**

More Mining Books

<u>Prospecting and Developing A Small Mine</u> - Topics covered include the classification of varying ores, how to take a proper ore sample, the proper reduction of ore samples, alluvial sampling, how to understand geology as it is applied to prospecting and mining, prospecting procedures, methods of ore treatment, the application of drilling and blasting in a small mine and other topics that the small scale miner will find of benefit. **8.5" X 11", 112 ppgs, Retail Price: $11.99**

<u>Timbering For Small Underground Mines</u> - Topics covered include the selection of caps and posts, the treatment of mine timbers, how to install mine timbers, repairing damaged timbers, use of drift supports, headboards, squeeze sets, ore chute construction, mine cribbing, square set timbering methods, the use of steel and concrete sets and other topics that the small underground miner will find of benefit. This volume also includes twenty eight illustrations depicting the proper construction of mine timbering and support systems that greatly enhance the practical usability of the information contained in this small book. **8.5" X 11", 88 ppgs. Retail Price: $10.99**

<u>Timbering and Mining</u> - A classic mining publication on Hard Rock Mining by W.H. Storms. Unavailable since 1909, this rare publication provides an in depth look at American methods of underground mine timbering and mining methods. Topics include the selection and preservation of mine timbers, drifting and drift sets, driving in running ground, structural steel in mine workings, timbering drifts in gravel mines, timbering methods for driving shafts, positioning drill holes in shafts, timbering stations at shafts, drainage, mining large ore bodies by means of open cuts or by the "Glory Hole" system, stoping out ore in flat or low lying veins, use of the "Caving System", stoping in swelling ground, how to stope out large ore bodies, Square Set timbering on the Comstock and its modifications by California miners, the construction of ore chutes, stoping ore bodies by use of the "Block System", how to work dangerous ground, information on the "Delprat System" of stoping without mine timbers, construction and use of headframes and much more. This volume provides a reference into not only practical methods of mining and timbering that may be employed in narrow vein mining by small miners today, but also rare insights into how mines were being worked at the turn of the 19th Century. **8.5" X 11", 288 ppgs. Retail Price: $24.99**

A Study of Ore Deposits For The Practical Miner - Mining historian Kerby Jackson introduces us to a classic mining publication on ore deposits by J.P. Wallace. First published in 1908, it has been unavailable for over a century. Included are important insights into the properties of minerals and their identification, on the occurrence and origin of gold, on gold alloys, insights into gold bearing sulfides such as pyrites and arsenopyrites, on gold bearing vanadium, gold and silver tellurides, lead and mercury tellurides, on silver ores, platinum and iridium, mercury ores, copper ores, lead ores, zinc ores, iron ores, chromium ores, manganese ores, nickel ores, tin ores, tungsten ores and others. Also included are facts regarding rock forming minerals, their composition and occurrences, on igneous, sedimentary, metamorphic and intrusive rocks, as well as how they are geologically disturbed by dikes, flows and faults, as well as the effects of these geologic actions and why they are important to the miner. Written specifically with the common miner and prospector in mind, the book will help to unlock the earth's hidden wealth for you and is written in a simple and concise language that anyone can understand. **8.5" X 11", 366 ppgs. Retail Price: $24.99**

Mine Drainage - Unavailable since 1896, this rare publication provides an in depth look at American methods of underground mine drainage and mining pump systems. This volume provides a reference into not only practical methods of mining drainage that may be employed in narrow vein mining by small miners today, but also rare insights into how mines were being worked at the turn of the 19th Century. **8.5" X 11", 218 ppgs. Retail Price: $24.99**

Fire Assaying Gold, Silver and Lead Ores - Unavailable since 1907, this important publication was originally published by the Mining and Scientific Press and was designed to introduce miners and prospectors of gold, silver and lead to the art of fire assaying. Topics include the fire assaying of ores and products containing gold, silver and lead; the sampling and preparation of ore for an assay; care of the assay office, assay furnaces; crucibles and scorifiers; assay balances; metallic ores; scorification assays; cupelling; parting' crucible assays, the roasting of ores and more. This classic provides a time honored method of assaying put forward in a clear, concise and easy to understand language that will make it a benefit to even beginners. **8.5" X 11", 96 ppgs. Retail Price: $11.99**

Methods of Mine Timbering - Originally published in 1896, this important publication on mining engineering has not been available for nearly a century. Included are rare insights into historical methods of timbering structural support that were used in underground metal mines during the California that still have a practical application for the small scale hardrock miner of today. **8.5" X 11", 94 ppgs. Retail Price: $10.99**

The Enrichment of Copper Sulfide Ores - First published in 1913, it has been unavailable for over a century. Topics include the definition and types of ore enrichment, the oxidation of copper ores, the precipitation of metallic sulfides. Also included are the results of dozens of lab experiments pertaining to the enrichment of sulfide ores that will be of interest to the practical hard rock mine operator in his efforts to release the metallic bounty from his mine's ore. **8.5" X 11", 92 ppgs. Retail Price: $9.99**

A Study of Magmatic Sulfide Ores - Unavailable since 1914, this rare publication provides an in depth look at magmatic sulfide ores. Some of the topics included are the definition and classification of magmatic ores, descriptions of some magmatic sulfide ore deposits known at the time of publication including copper and nickel bearing pyrrhotic ore bodies, chalcopyrite-bornite deposits, pyritic deposits, magnetite-ileminite deposits, chromite deposits and magmatic iron ore deposits. Also included are details on how to recognize these types of ore deposits while prospecting for valuable hardrock minerals. **8.5" X 11", 138 ppgs. Retail Price: $11.99**

The Cyanide Process of Gold Recovery - Unavailable since 1894 and released under the name "The Cyanide Process: Its Practical Application and Economical Results", this rare publication provides an in depth look at the early use of cyanide leaching for gold recovery from hardrock mine ores. This volume provides a reference into the early development and use of cyanide leaching to recover gold. **8.5" X 11", 162 ppgs. Retail Price: $14.99**

California Gold Milling Practices - Unavailable since 1895 and released under the name "California Gold Practices", this rare publication provides an in depth look at early methods of milling used to reduce gold ores in California during the late 19th century. This volume provides a reference into the early development and use of milling equipment during the earliest years of the California Gold Rush up to the age of the Industrial Revolution. Much of the information still applies today and will be of use to small scale miners engaging in hardrock mining. **8.5" X 11", 104 ppgs. Retail Price: $10.99**

Leaching Gold and Silver Ores With The Plattner and Kiss Processes - Mining historian Kerby Jackson introduces us to a classic mining publication on the evaluation and examination of mines and prospects by C.H. Aaron. First published in 1881, it has been unavailable for over a century and sheds important light on the leaching of gold and silver ores with the Plattner and Kiss processes. **8.5" X 11", 204 ppgs. Retail Price: $15.99**

The Metallurgy of Lead and the Desilverization of Base Bullion - First published in 1896, it has been unavailable for over a century and sheds important light on the the recovery of silver from lead based ores. Some of the topics include the properties of lead and some of its compounds, lead ores such as galenite, anglesite, cerussite and others, the distribution of lead ores throughout the United States and the sampling and assaying of lead ores. Also covered is the metallurgical treatment of lead ores, as well as the desilverization of lead by the Pattinson Process and the Parkes Process. Hofman's text has long been considered one of the most important early works on the recovery of silver from lead based ores. 8.5" X 11", 452 ppgs. Retail Price: $29.99

Ore Sampling For Small Scale Miners - First published in 1916, it has been unavailable for over a century and sheds important light on historic methods of ore sampling in hardrock mines. Topics include how to take correct ore samples and the conditions that affect sampling, such as their subdivision and uniformity. Particular detail is given to methods of hand sampling ore bodies by grab sample, pipe sample and coning, as well as sampling by mechanical methods. Also given are insights into the screening, drying and grinding processes to achieve the most consistent sample results and much more. 8.5" X 11", 124 ppgs. Retail Price: $12.99

The Extraction of Silver, Copper and Tin from Ores - First published in 1896, it has been unavailable for over a century and sheds important light on how historic miners recovered silver, copper and tin from their mining operations. The book is split into three sections, including a discussion on the Lixiviation of Silver Ores, the mining and treatment of copper ores as practiced at Tharsis, Spain and the smelting of tin as it was practiced by metallurgists at Pulo Brani, Singapore. Also included is an overview and analysis of these historic metal recovery methods that will be of benefit to those interested in the extraction of silver, copper and tin from small mines. 8.5" X 11", 118 ppgs. Retail Price: $14.99

The Roasting of Gold and Silver Ores - First published in 1880, it has been unavailable for over a century and sheds important light on how historic miners recovered gold and silver rom their mining operations. Topics include details on the most important silver and free milling gold ores, methods of desulphurization of ores, methods of deoxidation, the chlorination of ores, methods and details on roasting gold and silver ores, notes on furnaces and more. Also included are details on numerous methods of gold and silver recovery, including the Ottokar Hofman's Process, the Patera Process, Kiss Process, Augustin Process, Ziervogel Process and others. 8.5" X 11", 178 ppgs. Retail Price: $19.99

The Examination of Mines and Prospects - First published in 1912, it has been unavailable for over a century and sheds important light on how to examine and evaluate hardrock mines, prospects and lode mining claims. Sections include Mining Examinations, Structural Geology, Structural Features of Ore Deposits, Primary Ores and their Distribution, Types of Primary Ore Deposits, Primary Ore Shoots, The Primary Alteration of Wall Rocks, Alterations by Surface Agencies, Residual Ores and their Distribution, Secondary Ores and Ore Shoots and Vein Outcrops. This hard to find information is a must for those who are interested in owning a mine or who already own a lode mining claim and wish to succeed at quartz mining. 8.5" X 11", 250 ppgs. Retail Price: $19.99

Garnets: Their Mining, Milling and Utilization - First published in 1925, it has been unavailable since those days and sheds important light on the mining, milling and utilization of garnets. Included are details on the characteristics of garnets, where they are found and how they were mined. 78 ppgs, 10.99

Gemstones and Precious Stones of North America - Leading mining historian Kerby Jackson introduces us to a classic mining publication on the gems and precious stones of the United States, Canada and mexico. First published in 1890, it has been unavailable since those days and sheds important light on the gems and precious stones that may be found in North America. Included are chapters on diamonds, corundum, sapphire, ruby, topaz, emerald, disapore, spinel, turquoise, tourmaline, garnets, beyrl, peridot, zircon, quartz crystals, feldspars, pearls and many others. Included are details on where these gems and precious stones may be found throughout North America, as well as their characteristics. 360 ppgs, 24.99

Mining Camps and Mining Districts - First released in 1885 by Charles Howard Shinn under the title "Mining Camps: A Study in American Frontier Government", this publication offers a unique look at how early gold miners established their own forms of representative government during the California Gold Rush. Drawing on the the early mining codes of mideviel German miners in the Harz Mountains, on the mining customs of the Cornish tin miners and early Spanish mining laws introduced into California, the miners established the first governments in the American West. 340 ppgs, 24.99

BLM Field Handbook for Mineral Examiners - Leading mining historian Kerby Jackson introduces us to a classic mining publication on mine evaluation. First published in 1962, this work sheds important light on the techniques of BLM Mineral Examiners to perform validity on mining claims. 132 ppgs, 10.99

<u>**Six Months In The Gold Mines During The California Gold Rush**</u> - Unavailable since 1850, this important work is a first hand account of one "49'ers" personal experience during the great California Gold Rush, shedding important light on one of the most exciting periods in the history of not only California, but also the world. Compiled from journals written between 1847 and 1849 by E. Gould Buffum, a native of New York, "Six Months In The Gold Mines During The California Gold Rush" offers a rare look into the day to day lives of the people who came to California to work in her gold mines when the state was still a great frontier. **8.5" X 11", 290 ppgs. Retail Price: $19.99**

<u>**The Discovery of Gold in Australia**</u> - First published in 1852, it has been unavailable since those days and sheds important light on Australia's gold mining history. Included are rare communications between British agents and the British Crown when gold was first discovered in Australia in 1851. This rare text contains hard to find details on Australia's first mining camps and Britain's early attempts to provide for the orderly regulation of gold mines in that part of the world. Also of interest are hard to find extracts of articles that appeared in the early colonial newspapers that did their best to report on Australia's gold rush as it took place.
102 ppgs, 10.99